Smart Construction and Industry 5.0

This book addresses the profound changes facing the construction sector to provide a comprehensive and forward-looking exploration of how renewable energy systems and intelligent technologies are reshaping the construction industry towards a more sustainable, resilient, and cost-effective future.

The authors begin by examining the global development of solar, wind, and hydropower systems, emphasizing their potential to decarbonize the built environment when effectively integrated into building and infrastructure design. The discussion extends to the role of construction practices in advancing sustainability, showcasing real-world projects such as net-zero and energy-positive buildings that exemplify the application of renewable technologies in modern construction. The book explores the impact of Industry 4.0 and Construction 5.0, presenting how digital tools such as artificial intelligence, IoT, robotics, blockchain, and big data are transforming design processes, site operations, energy management, and performance monitoring. It also addresses how automation and human–robot collaboration are redefining productivity, safety, and workforce dynamics in construction. Central to the book is the integration of building information modelling as a collaborative platform for planning, simulating, and managing renewable energy facilities. It further introduces Life Cycle Cost Analysis as a critical decision-making framework, linking economic viability with environmental performance across a project's lifespan. The research-informed analysis and practical case examples provided throughout allow readers to implement intelligent and low-carbon solutions in the built environment.

As this book provides valuable insights on technical, financial, and energy resource issues, it is an invaluable resource for researchers and anyone in the construction industry who is seeking to improve sustainability, productivity, and efficiency in their projects. It is especially beneficial for architects and designers, engineers, project managers, construction managers, and contractors, among others.

Smart Construction and Industry 5.0

Renewable Energy Integration, Digitalization, Robotics, and Sustainability

Abdul Mateen Khan, Wesam Salah Alaloul,
Syed Muhammad Yasir Ashrafi, and Muhammad Ali Musarat

CRC Press
Taylor & Francis Group
Boca Raton London New York

CRC Press is an imprint of the
Taylor & Francis Group, an **informa** business

Designed cover image: Shutterstock

First edition published 2026
by CRC Press
2385 NW Executive Center Drive, Suite 320, Boca Raton FL 33431

and by CRC Press

4 Park Square, Milton Park, Abingdon, Oxon, OX14 4RN

CRC Press is an imprint of Taylor & Francis Group, LLC

ISBN: 9781032982274 (hbk)
ISBN: 9781032985497 (pbk)
ISBN: 9781003599272 (ebk)

DOI: 10.1201/9781003599272

Typeset in Palatino
by KnowledgeWorks Global Ltd.

Contents

Author Biographies

Abdul Mateen Khan is a Graduate Research Assistant at Universiti Teknologi PETRONAS, Malaysia. His research focuses on building information modelling (BIM), digital construction, safety, and intelligent monitoring systems. He specializes in integrating BIM with IoT and AI to enable real-time safety and performance management in construction environments. His work has contributed to BIM-based energy optimization, digital life cycle tracking, and ISO 19650–based frameworks. With over 825 citations, an h-index of 17, and an i10-index of 20, his research supports the advancement of AI-driven safety and decision-making systems in the construction sector.

Wesam Salah Alaloul is an Associate Professor in Civil and Environmental Engineering at Universiti Teknologi PETRONAS, Malaysia. His expertise includes digital construction technologies, life cycle costing, safety, and sustainability. With over 280 publications, an h-index of 47, and approximately 7,800 citations (as of mid-2025), his work has influenced areas such as AI-enabled safety systems, BIM integration, and automated project monitoring. He is an active contributor to research on the digitization of construction practices in emerging economies.

Syed Muhammad Yasir Ashrafi is a PhD candidate in Civil and Environmental Engineering at Universiti Teknologi PETRONAS, Malaysia. His research centres on using computer vision, particularly action recognition, pose estimation, and deep learning, to detect unsafe behaviours on construction sites. His goal is to develop intelligent systems that proactively enhance safety through real-time monitoring and automated hazard detection, contributing to AI-integrated safety frameworks in the construction industry.

Muhammad Ali Musarat is a Civil Engineer specializing in Construction Engineering and Management, with extensive experience in Research and Development. He has published numerous research articles and focuses his research on Digital and Automated Construction. In addition to his articles, Dr. Musarat is the author of several books and holds multiple patents in this field.

1

Introduction to Renewable Energy

1.1 Introduction

Energy lies at the very epicentre of human development and conventionally has moulded and shaped civilizations and advanced progress. Ever since the rise of human life, energy was needed to survive. The discovery of fire was one of the major milestones because it kept people warm and safe and gave food that could be cooked, improving health and furthering human life expectancy. It formed the very foundation of the growth of human civilization. With the development of societies came a further development in their needs for energy. Biomass – woody plants and animal dung – was the primary material early human settlements used to meet basic energy needs. These were, in fact, truly sustainable at that time because they were naturally replenished and caused minimal environmental impact. With the increase in population and the commencement of industries, there was further need for forms of energy that could be potent and efficient. Coal and, later, oil and natural gas discovered during the Industrial Revolution completely changed the world. These fossil fuels fed the factories, trains, and ships for economic prosperity and burgeoning technological development never seen before. Fossil fuels became the backbone of modern industrialized societies, providing the energy to build infrastructure, transport goods, and fuel growth in urban centres. This reliance on fossil fuels did not come without its significant consequences. Burning these fuels – principally coal, oil, and natural gas – releases huge volumes of carbon dioxide along with other noxious pollutants into the atmosphere, causing air pollution and contributing to the ominously changing climate. In addition, these resources are finite; in other words, they will eventually run out (Deshmukh et al., 2023; Sorensen, 2004).

This obviously raises some very serious questions over long-term energy security. By the end of the 20th century, it was already clear that dependence on fossil fuels could not be a future solution for the planet or its occupants either. This growing realization ignited the search for cleaner, more sustainable alternatives – what today we call renewable energy. Basically, renewable energy is energy that is generated from natural processes which

DOI: 10.1201/9781003599272-1

are continuously replenished, including sunlight, wind, water, and geothermal heat. Such energy resources, other than fossil fuels, do not run out, and their environmental impact is much lower. It means the total change of conception about the production and consumption of energy everywhere in the world. Such is the case with solar energy, which converts the power of the sun into electrical energy. Wind energy tends to capture the kinetic energy from the wind through turbines. Hydropower, in general, is produced using moving water for electricity generation, while geothermal energy is a kind of heat coming out of the interior of the Earth (Kothari et al., 2021; Sørensen, 1991).

These would appear to be the advantages of renewable energy: plentiful and sustainable sources, little or no greenhouse gas emissions, major solutions in the fight against climate change. Besides that, over time, renewable energy technologies became cheaper and more efficient, hence more competitive against traditional fossil fuels. At the forefront of all the ways the world strives for a cleaner, greener future is renewable energy. It especially seems to inspire governments, industries, and communities of every calibre to invest in renewable energy infrastructure, since such a resource could meet the future demand for energy requirements with due respect for environmental concerns. In fact, it is only expected that renewable energy will continue to play a stronger role in shaping a more sustainable and resilient world for future generations.

1.1.1 The History of Renewable Energy

Development and utilization of renewable energy sources are by no means modern-day phenomena, but date back to thousands of years in history. Early human civilizations utilized the boundless energy supply from nature to meet all their needs. It is such practices that formed the very foundation of what today is termed renewable energy, as discussed in Table 1.1.

1.1.1.1 Ancient Innovations

Wind energy is among the very oldest forms of renewable energy that have ever been harnessed by humans. As early as AD 500–900, there were reports of ancient Persians making windmills, so-called 'windmills of Persia', to grind grain, as shown in Figure 1.1. Large blades caught the force of the wind and turned them, thereby activating a milling mechanism. This innovation not only made food processing more efficient but also showed that humanity could have genius ideas in applying natural forces for practical ends.

- *Interesting Fact:* **The first recorded design of windmills appeared during the 9th century in Persia (now Iran), with vertical sails of cloth or reeds attached to a central post. That is how they manage to utilize the energy of the wind, even in low-wind areas.**

TABLE 1.1

Comprehensive Historical Evolution of Renewable Energy Technologies

Time Period	Civilization/ Region	Technology Innovation	Primary Application	Energy Output/ Efficiency	Notable Features	Environmental Impact	Societal Benefits	Key Innovators/ Pioneers
4000-3000 BCE	Ancient Egypt	Sail boats on Nile	Transportation and trade	Wind-dependent; 3–5 knots average speed	Reed sails; wooden hulls	Zero emissions	Enabled long-distance trade	Unknown Egyptian sailors
200 BCE	Ancient Greece	Water clocks (Clepsydras)	Time measurement and irrigation	Gravity-fed; 95% efficiency	Bronze/stone construction	Minimal water usage	Accurate time keeping	Ctesibius of Alexandria
500-900 CE	Persian Empire	Vertical-axis windmills	Grain grinding and water pumping	15–20% efficiency	Cloth/reed sails; stone base	Zero emissions	Agricultural productivity	Mathematician Al-Farabi
1000-1200 CE	China	Horizontal-axis windmills	Water pumping and irrigation	20–30% efficiency	Bamboo construction	Biodegradable materials	Enhanced agriculture	Unknown Chinese engineers
1300-1500 CE	Netherlands	Dutch windmills	Land drainage and lumber cutting	25–35% efficiency	Four-blade design; rotating cap	Sustainable materials	Land reclamation	Cornelis Corneliszoon
1767 CE	Switzerland	Hot box solar collector	Heat trapping and study	30–40% heat retention	Glass panels; insulation	Zero emissions	Scientific advancement	Horace de Saussure
1860s CE	France	Solar steam engine	Industrial power	3–5% efficiency	Parabolic reflectors	Carbon-neutral	Industrial innovation	Augustin Mouchot
1880s CE	United States	Selenium photovoltaic cells	Electricity generation	1–2% efficiency	Selenium wafers	Minimal waste	Scientific breakthrough	Charles Fritts

(Continued)

TABLE 1.1 ***(Continued)***

Comprehensive Historical Evolution of Renewable Energy Technologies

Time Period	Civilization/ Region	Technology Innovation	Primary Application	Energy Output/ Efficiency	Notable Features	Environmental Impact	Societal Benefits	Key Innovators/ Pioneers
1930s CE	United States	Smith-Putnam wind turbine	Grid electricity	1.25 MW capacity	Steel construction	Low noise pollution	Grid-scale power	Palmer Putnam
1970s CE	Global	Silicon solar cells	Commercial power	10–15% efficiency	Silicon wafer technology	Manufacturing impact	Energy independence	Bell Labs scientists
1990s CE	Denmark	Offshore wind farms	Grid electricity	20–30% efficiency	Marine-grade materials	Marine ecosystem impact	Large-scale clean energy	Bonus Energy A/S
2000s CE	Germany	Advanced photovoltaics	Distributed generation	15–20% efficiency	Thin-film technology	Reduced carbon footprint	Energy democratization	Multiple manufacturers
2010s CE	China	Ultra-efficient solar	Utility power	20–25% efficiency	Perovskite materials	Recycling challenges	Cost-effective power	Multiple research teams
2020s CE	Global	Floating offshore wind	Deep-sea power generation	35–45% efficiency	Floating foundations	Minimal seabed impact	Ocean energy capture	Multiple developers

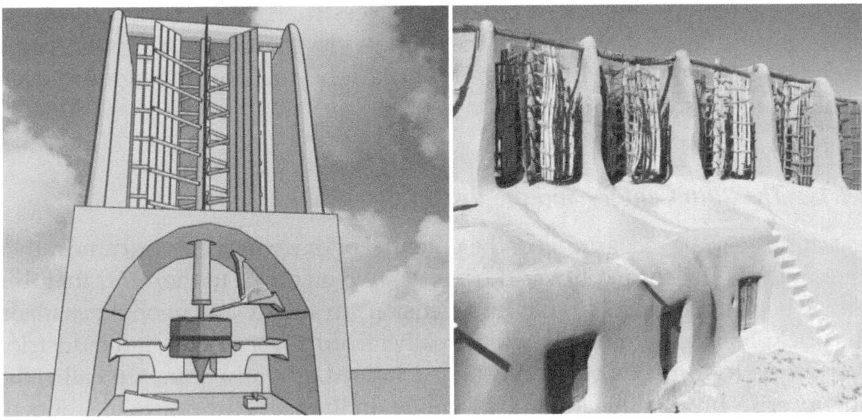

FIGURE 1.1
Vertical shaft windmills (Hills, 1996).

In addition to wind, ancient Greeks and Romans harnessed kinetic energy from flowing water by employing water wheels. These water wheels converted the flow of rivers and streams into mechanical energy, enabling the milling of grain, sawing wood, and even forging. Water wheels were an important factor in the development of early industry and agriculture through the illustration of renewable energy sources well before the modern era.

Example:

A version of the water wheel, called Noria, was developed in ancient times for irrigation in Syria and Egypt. It was a large wheel with buckets that lifted water from a river for irrigation – an early application of hydropower.

1.1.1.2 The Shift to Industrialization

Despite these early innovations, the systematic development of renewable energy technology took hold a great deal later. The Industrial Revolution in the 18th and 19th centuries saw a severe break in trends in the use of energy. The wide use of coal followed by oil and natural gas supported the rapid expansion of fossil-fuel-driven industries. This age was typified by little else but this preoccupation with such abundant energy resources, throwing renewable energy resources into relative obscurity. Among the major breakthroughs of renewable energy in the 19th century was the invention of the first solar-powered steam engine, credited to French scientist Augustin Mouchot in the 1860s. Mouchot's invention showed that solar radiation could be viable for producing motive power in other forms, for example, by generating steam. This innovation, with immense promise, was not taken up within a very short period due to the ample availability of fossil fuels at a very low cost.

While renewable resources contributed less than 20%, at the end of the 19th century, more than 80% of the worldwide energy consumption came from fossil fuels, which ultimately reduced renewable technologies to even smaller portions (Malanima, 2014).

1.1.1.3 The 20th Century and the Rise of Renewables

In fact, large-scale recognition of the potential of renewable energy came rather later in the 20th century. It was during the two oil crises in the 1970s that this slowly began to change. In that time, tensions in supply and geopolitics made the price of oil shoot upwards, which really forced governments and industries worldwide to think twice about energy. Indeed, it was a wake-up call – the energy crisis brought to light vulnerabilities in the economy due to overdependence on fossil fuels. It is in view of this that a sense of urgency made countries seriously invest in various research and development in renewable energy. In due course, Denmark became the pacesetter for wind energy, which translated favourable geography into handsome banks of wind turbines. Towards the end of the 20th century, Denmark produced a tenth of its electricity through the wind – a global benchmark on integrating renewable energy.

By 1990, more than 2,000 MW of wind capacity had been installed in Denmark, supplying about a quarter of the nation's electricity needs – and still climbing (MacKenzie, 2021).

Meanwhile, Germany was playing the most decisive role in the development of renewable energy. Several supportive policies adopted by the German government – especially for the growth of renewable technologies – advanced solar energy. Following feed-in tariffs introduced in the early 2000s, building photovoltaic systems incentivized such a revolution to let households and even businesses generate their own electricity and sell excess power back to the grid.

- *Interesting Fact:* **Already in 2012, Germany became a world leader in solar energy, and solar power in this country accounted for about 26% of the nation's all-electric output during especially sunny days.**

In the United States, attention was now focused on renewable energies, primality energy security, and environmental sustainability. Many research institutions and private enterprises were investing in solar and wind technologies, hence dramatic improvements and cost reductions for the technologies.

Example:

The California Solar Initiative started in 2006 aimed at installing 1,940 MW of new solar capacity until 2016 and to make California the leading solar market.

Gaining considerable momentum at the turn of the 21st century, renewable energy had been spurred on by technological advances during the late 20th century, along with a growing public awareness over climate change and environmental issues. The SET plan sets ambitious targets for the deployment of renewable energy to cut down greenhouse gas emissions, placing countries on a pathway towards sustainable energy systems.

Renewables, driven by wind and solar, accounted for about 29% of all electricity generated in the world last year. In 2022, solar power accounted for more than 12% of the world's electricity, according to the International Energy Agency (Valavanidis, n.d.). Today, renewable energies are no quixotic alternative; they are instead a keystone to a secure energy future. Determination against climate change is growing with continuous technological improvement, while so is the promising history of renewable energy for a cleaner and resilient world that future generations can inherit.

THOUGHT BOX

- Early societies were already masters of renewable energy, using wind and water creatively to power daily life.
- Fossil fuels became dominant during the Industrial Revolution, but their environmental impact has prompted a return to renewables.
- By the 1970s, global energy crises highlighted the risks of fossil fuel dependence, sparking innovation in wind and solar energy.

QUIZ

1. Which ancient tool used wind power to grind grain?
2. Which major event in the 1970s revived global interest in renewables?
3. Name one country that became a leader in wind energy by the late 20th century.

1.1.2 Renewable Energy Growth: Milestones and Indicators

In the past few decades, the renewable energy sector has literally seen exponential growth and has irrevocably altered the paradigm of the world energy configuration. The rise is not a temporary tendency but a fundamental transition in the way the world generates power. With worsening manifestations of climate change through extreme weather patterns, rising sea levels, and environmental degradation, the necessity to adopt cleaner, sustainable sources of energy has brought the renewable energy sector into the current focus of worldwide energy deliberations.

1.1.2.1 Global Growth Trends

The International Renewable Energy Agency (IRENA) estimated that the global renewable energy capacity grew from approximately 1,200 GW back in 2010 to over 3,000 GW by the year 2020 (Allegretti et al., 2024), as shown in Figure 1.2. Good growth in this sector can be blamed on several related factors:

1. *Falling Costs:* The costs related to the technologies of renewable energy have fallen considerably in recent years. For example, it has been indicated that on average, solar photovoltaic (PV) systems are about 89% cheaper than they were back in 2010 (Gul et al., 2016). Such a shift has really given most people an opportunity to access the use of solar energy. Onshore wind power has also become cheaper by as much as 70% over the same period. These reductions may be partly because of enhancements in manufacturing processes, improved efficiency of technology, and greater competition within the energy market. Falling costs have now enabled countries and companies

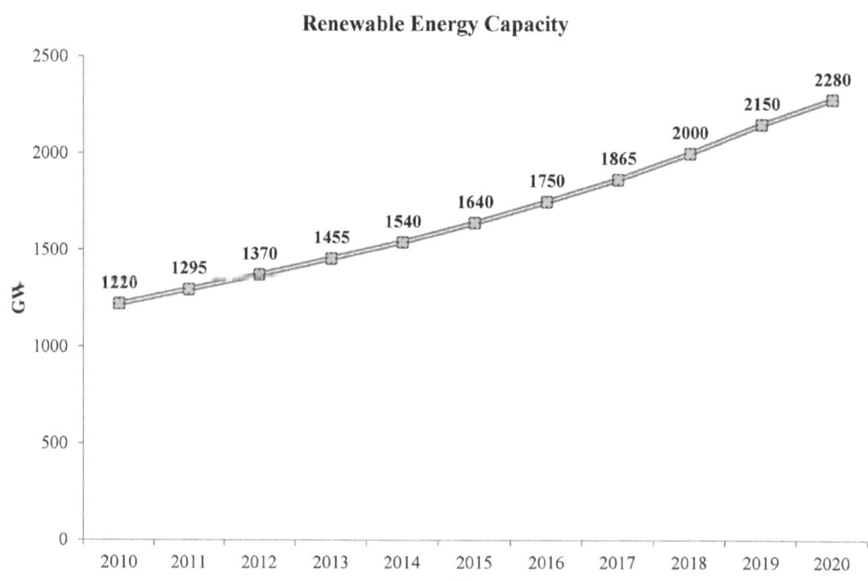

FIGURE 1.2
Global renewable energy growth (2010–2020) (Mulopo, 2022).

to invest in those renewable projects which were not economically viable in the past.

2. *Technological Innovation:* There has been a considerable reduction in the cost of renewable technologies. This has increased the efficiency and reliability of various renewable sources of energy. Innovation in different energy storage solutions, such as lithium-ion batteries, enables the storage of energy produced from intermittent sources like solar and wind, as shown in Figure 1.3. Secondly, the development of smart grid mechanisms has immensely enhanced the integration of renewable energies within conventional power systems for better supply and demand management. This sort of technological improvement not only optimizes energy use but also promotes overall reliability in renewable energy systems.
3. *Climate Change Awareness:* The increasing awareness about the issue of climate change and its consequences has driven the entire world towards cleaner sources of energy. The realization of the need to reduce greenhouse gas emissions has set a supportive

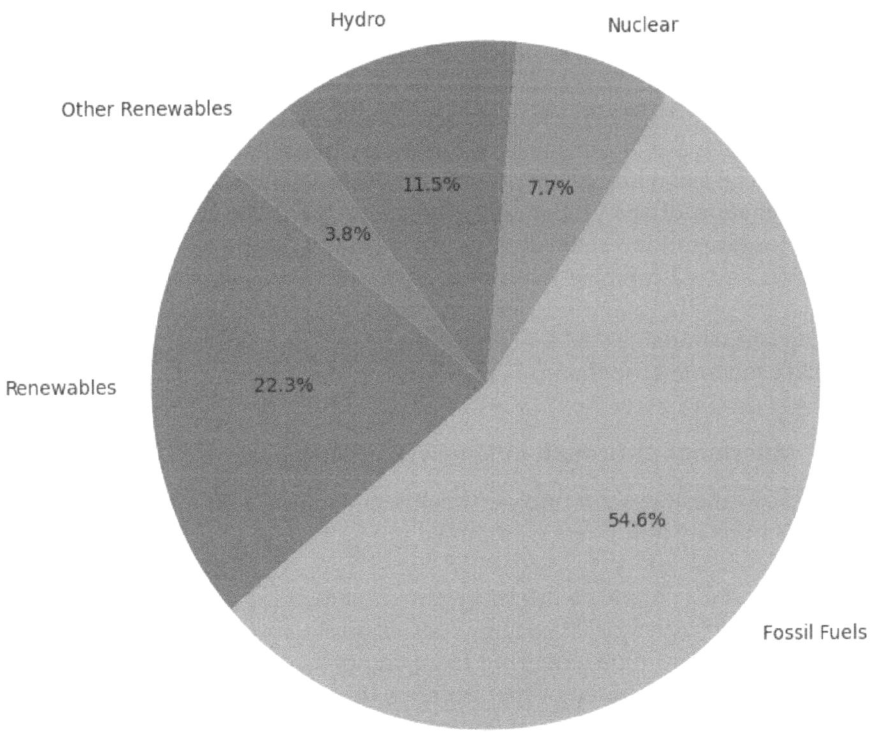

FIGURE 1.3
Global electricity generation from renewable energy (Hasanuzzaman et al., 2017).

policy environment and stirred investment into renewable energy. Ambitious targets for carbon reduction and increasing the percentage share of renewable supplies in the energy mix have been set by several countries. This change has also been accompanied by the growth of grassroots movements to advocate for sustainable energy solutions.

THOUGHT BOX

WHAT IF FOSSIL FUELS DISAPPEARED TOMORROW?

Imagine a sudden disappearance of all fossil fuel reserves. How would global infrastructure, economies, and societies adapt overnight?

This growth is being measured not only in capacity but also in its significant impact on global electricity generation and greenhouse gas emissions. A development that reflects a gradual transition from fossil fuels towards its more sustainable alternative, renewable energy sources, for instance, accounts for about 29% of global electricity generation in 2022. The shift has resulted in a notable decrease in carbon emissions; for example, according to IRENA, renewable energy avoided nearly 2.1 gigatonnes of carbon dioxide emissions in 2020 alone (Asmelash & Gorini, 2021). The International Renewable Energy Agency (IRENA) serves as the principal platform for international co-operation, a centre of excellence, a repository of policy, technology, resource and financial knowledge, and a driver of action on the ground to advance the transformation of the global energy system. An intergovernmental organisation organization established in 2011, IRENA promotes the widespread adoption and sustainable use of all forms of renewable energy, including bioenergy, geothermal, hydropower, ocean, solar and wind energy, in the pursuit of sustainable development, energy access, energy security, and low-carbon economic growth and prosperity.

1.1.3 Milestones of Growth in Renewable Energy

A few key milestones put into perspective the rapid evolution of these recent decades in renewable energy, as shown in Figure 1.4:

- *Early 2000s:* Many countries began implementing policies to develop renewable energies. Germany, for example, introduced the Feed-in Tariff programme, ensuring fixed compensation to produce renewable energies. Among other policies, this greatly spurred the development of solar and wind energy.
- *2010:* IRENA, was created to help the world realize the widespread adoption of renewable energy sources. Its establishment marked the

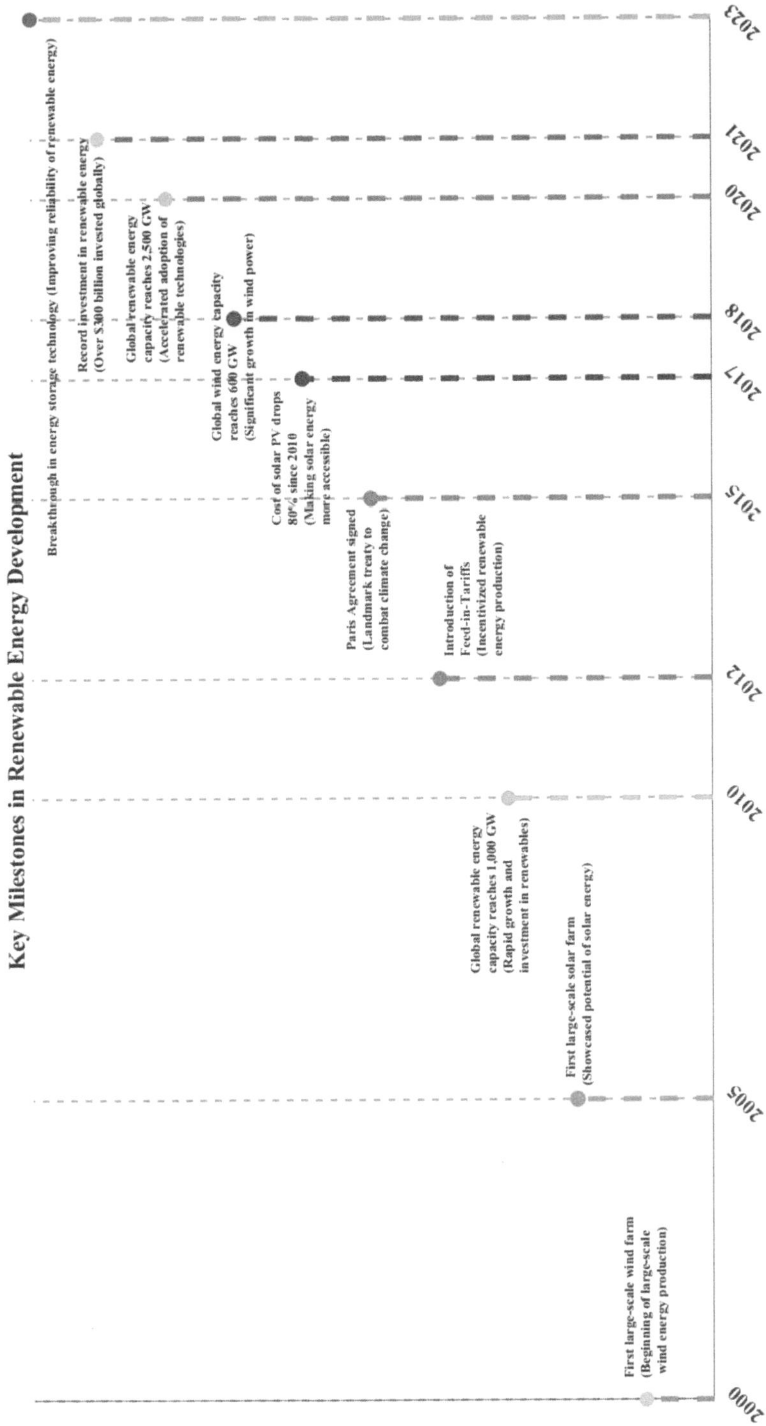

FIGURE 1.4
Key milestones in renewable energy development (Nwokediegwu et al., 2024).

increasing awareness by the global energy policy thinkers of the role of renewables.

- *2015:* The Paris Agreement was reached among almost 200 nations with an agreed framework on ways through which climate change can be combatted. There were calls for drastic cuts in greenhouse gas emissions, which further firmed up investments in renewable energy.
- *2020:* Due to the COVID-19 pandemic, energy demand declined temporarily worldwide. Yet this crisis emphasized and urged for more resilient and sustainable systems of energy that shall support development in the long term. Renewable energy continued to surge upwards even while economic activities are at a low. According to IRENA, renewables accounted for about 90% of new electricity generation worldwide.

1.1.3.1 Regional Developments

Attention to renewable energy does not spread uniformly around the world. Different regions have been experiencing different rates of development according to their local resources, government policy, and economic conditions.

- *Asia:* It has led the world in the deployment of renewable energy, mostly led by China. More than half of the world's solar capacity and a sizeable portion of its wind power capacity were accounted for by the nation in 2022. It has become a leader in renewable energy, thanks to its strict policies and investments in innovative technology.
- *Europe:* European countries have also led the way in the adoption of renewable energy sources. Countries such as Germany, Denmark, and Spain have invested heavily in funding for wind and solar energy; thus, these two sources comprise a high percentage share of their respective energy mixes, and this can be seen in Figure 1.5. For instance, in 2022, Denmark managed to produce approximately 47% of its electricity from wind power; this becomes a real-world case study to demonstrate the feasibility of large-scale integration of wind energy.
- *North America:* The United States has seen very rapid growth in deployment in both solar and wind energies. Investment has been facilitated using tax incentives and state-level renewable portfolio standards. By 2022, the United States had installed more than 130 GW of capacity, with California in the leading role. Texas has dominated the production of wind energy among other states.

As renewable energy technologies continue to evolve, understanding their technical and operational characteristics becomes crucial for informed

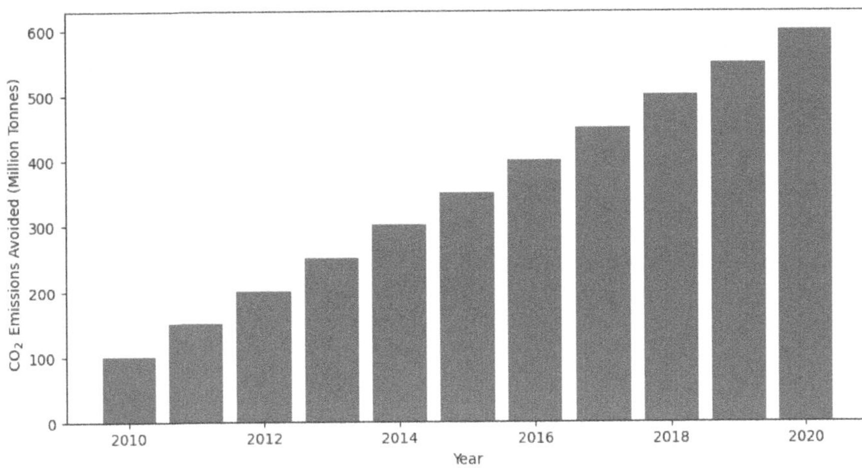

FIGURE 1.5
CO_2 emissions avoided annually (2010–2020) (Antonakakis & Usikov, 2024).

decision-making in energy planning. The following comprehensive comparison presents key metrics across various renewable technologies, focusing on practical aspects such as installation requirements, operational parameters, and integration challenges. This data establishes a baseline for comparing different renewable energy options beyond their basic power generation capabilities, helping stakeholders evaluate the full spectrum of considerations in renewable energy deployment, as shown in Table 1.2. The metrics presented are industry averages and may vary based on specific project conditions, technological advances, and local factors.

THOUGHT BOX

RENEWABLES AND CLIMATE JUSTICE

How can renewable energy support global equity? Consider how access, affordability, and implementation vary across different regions.

1.1.4 Solar Energy: A Beacon of Growth

Solar energy, captured in photovoltaic cells, has grown to be one of the major and fastest-growing sources of renewable energy on the global scene. This growth has represented not only technological advances and decreased costs but also a general shift in the world towards more sustainable energy solutions, as shown in Figure 1.6. In 2010, the world's cumulative installed solar capacity was about 40 GW; in 2022 it mushroomed to over 1,000 GW – more than a 2,400% increase in little more than a decade. This has made solar

TABLE 1.2
Comparative Analysis of Modern Renewable Energy Technologies (2024)

Technology Type	Average Lifespan (Years)	Installation Time	Space Requirement (per MW)	Initial Setup Cost ($/kW)	Maintenance Cost ($/kW/year)	Energy Storage Potential	Grid Integration Complexity	Noise Level (dB)	Raw Material Requirements	Manufacturing Complexity
Solar PV (mono-crystalline)	25-30	2–3 months	4–5 acres	800–1,000	15–20	Limited without batteries	Moderate	<40	High silicon purity	High
Solar PV (thin film)	20–25	1–2 months	5–5 acres	600–800	10–15	Limited without batteries	Moderate	<40	Varied semiconductors	Moderate
Onshore wind	20–25	6–12 months	40–70 acres	1,200–1,400	40–50	Not directly possible	Complex	45–60	Steel, composite materials	High
Offshore wind	25–30	12–24 months	80–100 acres	2,500–3,000	75–100	Not directly possible	Very complex	40–50	Corrosion-resistant materials	Very High
Small hydro (<10 MW)	50–100	12–18 months	Site-dependent	1,500–2,500	30–40	Natural storage	Simple	70–90	Concrete, steel	Moderate
Geothermal	30–50	24–36 months	1–8 acres	2,000–4,000	100–150	Constant output	Simple	65–85	Specialized drilling equipment	High
Biomass	20–25	18–24 months	3–4 acres	3,000–4,000	120–150	Fuel can be stored	Simple	60–80	Organic matter	Moderate
Tidal	20–25	36–48 months	Site-dependent	3,000–5,000	100–200	Predictable cycles	Complex	40–60	Marine-grade materials	Very high
Wave energy	15–20	24–36 months	Site-dependent	4,000–6,000	150–200	Not directly possible	Complex	40–50	Marine-grade materials	High
Concentrated solar power	25–30	18–24 months	5–10 acres	5,000–7,000	60–80	Thermal storage possible	Moderate	50–60	Mirrors, heat transfer fluid	High

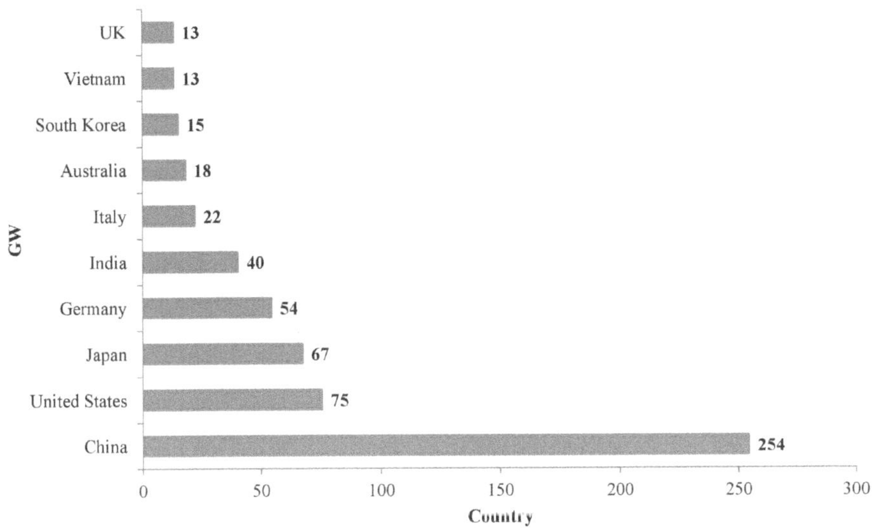

FIGURE 1.6
Top ten countries of solar energy development (Mahendru et al., 2024).

power one of the most promising sectors in the fight against climate change and an indispensable part of the global energy transition (Fessler, 2019).

1.1.4.1 Top Ten Countries of Solar Energy Development

More than a few countries have played pivotal roles in the proliferation of solar energy. Firstly, there is China, fast emerging as the world's largest solar market, contributing over 50% to the global solar capacity by 2022. The Chinese government instituted a raft of ambitious policies and incentives that drive solar adoption. This includes the 'Solar Roofs' programme, which incentivizes residential and commercial buildings to install solar panels. This has meant a sharp increase in installations across the nation. Secondly, China has heavily invested in the production of solar panels locally; this has reduced the costs and increased efficiency. By 2022, China had installed more than 400 GW of solar capacity and dominated the global market, setting a good example for other nations.

The United States comes second globally as per the installed solar capacity because of several incentives at both the federal and state levels. Perhaps the most important catalyst for this growth has been the ITC, which allows individuals and businesses to write off a significant percentage of their solar installations' cost from federal taxes. The United States has built up an installed solar capacity of approximately 130 GW by 2022, driven mostly by states such as California, Texas, and Florida. California leads the nation in solar adoption, driven by policies encouraging rooftop installations and utility-scale solar farms. In 2020, California generated 20% of its electricity from solar power – a real testament to how well the state's renewable energy policy is working. India has done an extremely wonderful job with its solar energy. For one, it shifted an ambitious target of 100 GW of solar capacity by the year 2022 under the National Solar Mission. The Indian government has implemented policies such as the setting up of solar parks, thereby facilitating large-scale solar installations. By 2022, India has achieved around 60 GW installed solar capacity, showing its determination for renewable energy. The government has also initiated programmes to promote solar energy in rural areas, with the ambition to provide electricity to millions of people who do not have access to reliable power.

THOUGHT BOX

THE MYTH OF CLEAN ENERGY?

Is renewable energy truly clean? Reflect on environmental costs related to production, land use, and material sourcing.

1.1.4.2 Notable Projects and Innovations

The Mohammed bin Rashid Al Maktoum Solar Park happens to be one of the ambitious solar projects across the globe, which is under implementation in Dubai. It indeed will become the world's largest single-site solar park with a projected capacity of 5,000 MW by 2030. Based on both PV and concentrated solar power technologies, the park therefore shows great potential for solar energy to be used in arid regions where plentiful sunlight has been received. Projects of such a large scale satisfy not only the local needs but also serve as an example for other countries participating in the race of solar energy application.

1.1.4.2.1 Case Study: The Topaz Solar Farm

The Topaz Solar Farm is a large solar farm with the capacity of 550 MW, located in San Luis Obispo County, California, as shown in Figure 1.7. It was commissioned in 2014 and has ever since continued to make strong contributions to the energy sector within the region. The plant consists of over 9 million solar panels, capable of generating enough electricity for about 160,000 homes in a year. Another excellent aspect of Topaz involves the contribution it makes to lowering the rate of greenhouse gas emissions. Because fossil-fuel-based power generation has been replaced, the solar farm can prevent approximately 600,000 metric tonnes of carbon dioxide emissions per year – akin to taking around 120,000 cars off the road. The project underlines how land use becomes highly significant in the development of solar energy. The Topaz Solar Farm was built on already disturbed land to reduce the environmental effect and protect natural habitats of species.

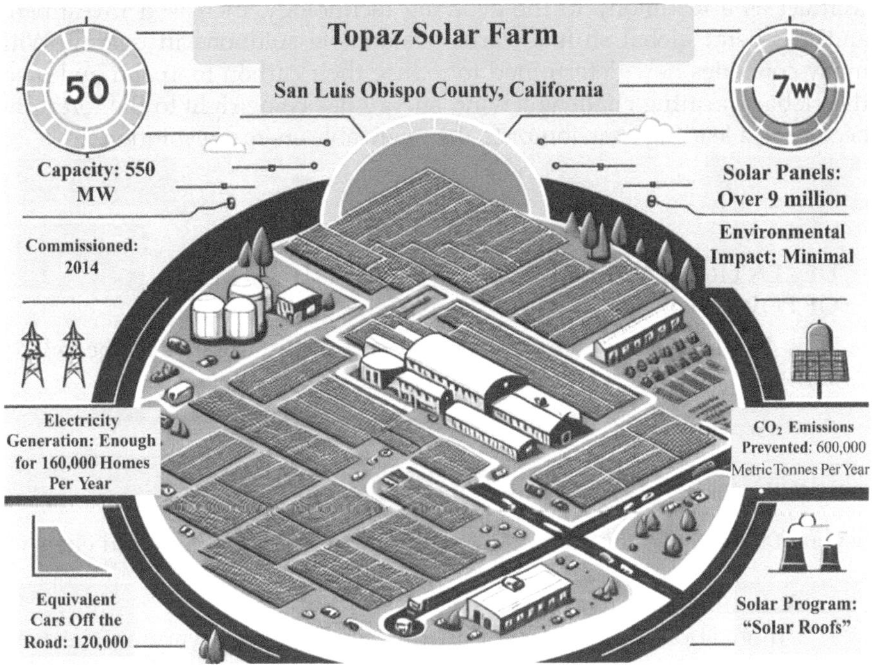

FIGURE 1.7
Topaz Solar farm in San Luis Obispo County, California.

1.1.4.3 Economic and Environmental Benefits

Solar growth has come with a plethora of economic and environmental benefits. New job creation occurs during manufacturing, installation, and maintenance, at each turn of the solar industry's growth. In fact, according to the National Solar Jobs Census by the Solar Foundation, solar employed over 250,000 workers in the United States solar market in 2020, representing a 167% growth in employment since 2010 (Regan et al., 2021). This growth of solar energy has also stimulated the local economies by allowing greater investments and increasing tax revenues. This means that in environmental terms, solar energy takes the place of a considerable amount of fossil fuel burning, which in turn reduces carbon emissions and increases air quality. Where countries worldwide are shifting towards solar power, the global dependence on coal and natural gas decreases, to which international efforts to fight against climate change are positively affected.

1.1.5 Wind Energy: Tapping into Nature

Over the years, wind energy has emerged as one of the fastest-growing sources of renewable energy; it recorded remarkable progress from 180 GW in 2010 to almost 800 GW by 2022 (Council, 2020). This surprise growth may well act as a testimony to the evolving technology, increased investment, and a general global shift towards sustainable solutions in energy. With many countries now determined to reduce their carbon footprint and meet the global warming challenge, wind energy has come right to the forefront, becoming a leading contributor to the renewable energy revolution.

THOUGHT BOX

DECENTRALIZED ENERGY SYSTEMS: THE FUTURE OF POWER?

How would society change if homes and communities became self-sufficient using solar panels and storage systems?

1.1.5.1 Key Drivers behind Growth in Wind Energy

Several factors have contributed to the exponential growth in wind energy:

1. **Technological Advances**
   Innovations in the design of turbines have been very crucial to making wind energy more efficient and effective. These new wind turbines had much larger rotor diameters and taller towers, which enabled them to catch more wind and convert it at a good, efficient rate. The latest onshore wind turbines can reach up to 200 m in height, further increasing their possibility of energy capture. Furthermore, studies in material science have successfully produced lighter and stronger turbine blades that can withstand higher conditions of wind and operate at higher capacities.
2. **Global Investment**
   The wind energy sector has found its consolidated ground for drawing huge investments, which indicates growing confidence in the sector's long term viability. In 2020 alone, over $20 billion was invested in the US wind energy sector, while state renewable energy mandates and federal tax incentives like Production Tax Credit (PTC) and Investment Tax Credit (ITC) acted as drivers. These financial incentives encourage new wind projects to be developed, contributing to the general reduction of energy costs. The total investment in wind energy around the world reached approximately US$140 billion in 2020, underlining its importance to the energy transition. Until 2022, Denmark has been regarded as the frontrunner in wind energy, with an estimated 47% of its total electricity consumption coming from this source. The Danish government has invested quite heavily in both onshore and offshore wind farms, placing it among the first-ranking countries in the whole world regarding

wind energy technology. The Hornsea Project in the North Sea is expected to yield 6,000 MW when completed. It is, indeed, an ambitious project that reflects upon Denmark's commitment to attaining carbon neutrality and hopefully will serve as a model for other countries to emulate. Germany has also managed to do some impressive feats in wind energy capture with the help of its Energiewende or energy transition policy, which aimed at taking the country towards a sustainable energy system. Germany has the target of getting 65% of its electrical generation contributed by renewable energy by 2030. In 2022, it has topped the ranking in the installed wind capacity onshore and offshore. The installed wind infrastructure is at a greatly robust level, thus becoming the world knowledge base for research and development with a high priority on turbine technology and grid integration.

Until 2022, China has attained more than 300 GW of installed capacity, reaching the status of the most prospective and leading producer of wind energy globally. The unparalleled growth of wind energy is because the government has especially supportive policies for the development of renewable energy in the country, coupled with significant on-site and offshore investments. The Gansu Wind Farm is one such project – one of the largest wind farms in the world, featuring 8,000 MW of capacity and an area larger than Portugal. This giant project will be able to generate enough electricity for about 8 million homes – a surefire way of showing the world how serious China is about renewable energy.

1.1.5.1.1 Gansu Wind Farm-Facts

Gansu Wind Farm, one of the world's biggest wind farms, is located in the Gansu province of China. Further, it has a surprising capacity of 8,000 MW and is on a land area larger than the whole of Portugal. This should give an example of the scale at which wind energy can be developed to meet enormous energy demands. The Gansu Wind Farm, upon completion, is expected to supply approximately enough amounts of electricity to power approximately 8 million homes, decidedly making huge contributions towards China's renewable energy goals and a decline in fossil fuel use.

1.1.5.1.2 Case Study: Walney Extension Offshore Wind Farm

The Walney Extension Offshore Wind Farm, commissioned in 2018, is located off the coast of Cumbria in the United Kingdom, as shown in Figure 1.8. This is one of the largest in the world with a capacity of 659 MW. This project will illustrate on a large scale the huge potential for large-scale offshore wind to provide for significant energy demands. The Walney Extension comprises 87 turbines, which can power over 600,000 homes throughout a year, thus being a significant enhancement to the UK's renewable energy targets.

FIGURE 1.8
(a) Walney Extension Offshore Wind Farm. (b) Walney Extension map.

Besides securing the supply of electricity, the project is very important for a low-carbon economy in that it has a vital role of contributing to the reduction of greenhouse gas emissions by generating clean electricity from the wind, enabling Walney Extension to displace fossil-fuel-fired generation, further cutting carbon dioxide and improving air quality.

THOUGHT BOX

WHAT WILL POWER THE NEXT INDUSTRIAL REVOLUTION?

If the 20th century was powered by oil and coal, what energy source will define the 21st century and why?

1.1.5.2 Economic and Environmental Benefits

The development of wind energy has given significant economic and environmental dividends. In the process, the wind energy industry is attracting many jobs for manufacturing, installation, and maintenance. According to the Global Wind Energy Council, the number employed in the wind industry was more than 1.4 million people all over the world in 2020, an increase of 20% as compared with the previous year (Council, 2020). These jobs are essentially important for the economic locals and make a significant contribution to attaining sustainable development. From an ecological perspective, wind energy serves to lessen dependence on burning fossil fuels, which by implication reduces the amount of carbon dioxide emitted, improving air quality in turn. In fact, according to a study conducted by the National Renewable Energy Laboratory, in 2019 alone, wind energy was forecast to reduce roughly 329 million metric tonnes of carbon dioxide emissions in the United States – a quantity which would correspond approximately to taking some 70 million cars off the road (Dykes et al., 2019).

1.1.6 Hydropower: Oldest Renewable

Energy Resource Hydropower is one of the oldest, most mature renewable energy resources, serving as it does the process of conversion of moving or falling water to electrical energy. It exploits this resource in terms of the gravitational force of the water through a conversion process from its potential energy to kinetic energy which is then converted into electrical energy. Hydropower accounts for approximately 16% of all generated electricity around the world in 2022 and is expected to continue playing a major role both in the transition into sustainable energy solutions and in mitigation efforts concerning climate change, as shown in Figure 1.9.

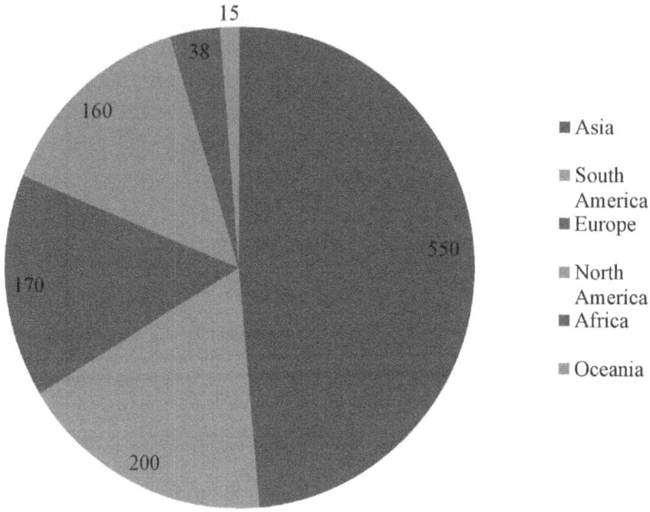

FIGURE 1.9
Energy resource hydropower in different continents (Wasti et al., 2022).

1.1.6.1 Global Capacity and Key Players

The total installed hydropower capacity worldwide grew steadily, reaching beyond 1,400 GW by 2022. This capacity has been contributed significantly by several countries, as shown in Figure 1.10.

1. **China**

   Presently, China holds the record of having the largest hydropower capacity in the world, with a huge network of hydroelectric plants across the country's borders. Among them, the major example of its type is the Three Gorges Dam on the Yangtze River which possesses a capacity of 22,500 MW. This huge structure provides enormous amounts of electricity to millions of households; it also contributes to the control of flooding and the assurance of river navigation. The dam has come to be a real symbol of the engineering prowess of China and the commitment the country attaches to renewable sources of energy.

2. **Brazil**

   Second in the world in hydropower generation, about 60% of the electricity produced in Brazil comes from hydro sources. The height of this dominance is reflected by the Itaipu Hydroelectric Dam along the border between Brazil and Paraguay, which has an installed capacity of 14,000 MW. The structure is quite pivotal in powering millions of people in Brazil and Paraguay, an indication of how important hydropower is within the energy landscape of the region.

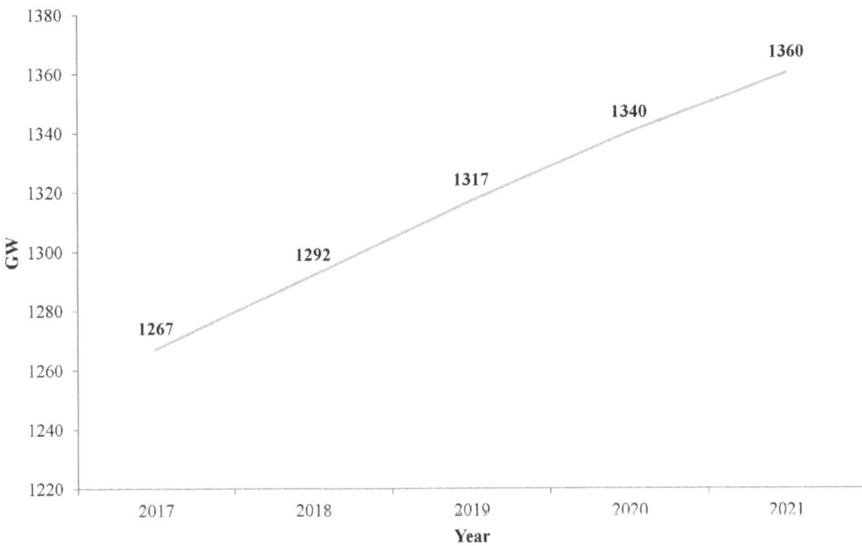

FIGURE 1.10
Hydropower installed capacity growth 2017–2021 (GW) (Sutabutr, 2012).

3. **The United States**
   This country also has significant hydropower potential, with its plants supplying about 7% of the country's electricity. The largest hydroelectricity power plant in the United States is the Grand Coulee Dam; it is situated in the state of Washington, with an installed capacity of 6,809 MW. It reflects the status of hydropower in the American energy mix and refers to another source of stable, renewable electricity.
4. **Canada**
   Another of the leaders in hydropower is Canada, where the equivalent of about 60% of the country's electricity is produced by hydroelectricity. The HydroQuébec system is among the largest in the world, and its facilities include the La Grande Complex in northern Québec, which supplies much of the province's energy.

1.1.6.2 Reliability and Baseload Power

One of the huge advantages that hydropower will provide is much more baseload power. In other words, it can supply electricity constantly and steadily, which is so important for grid reliability, as shown in Figure 1.11. This advantage comes against other sorts of renewable energy sources, such as solar and wind, which are intermittent in nature due to dependency on weather conditions; this makes hydropower operate day in, day out. Such reliability

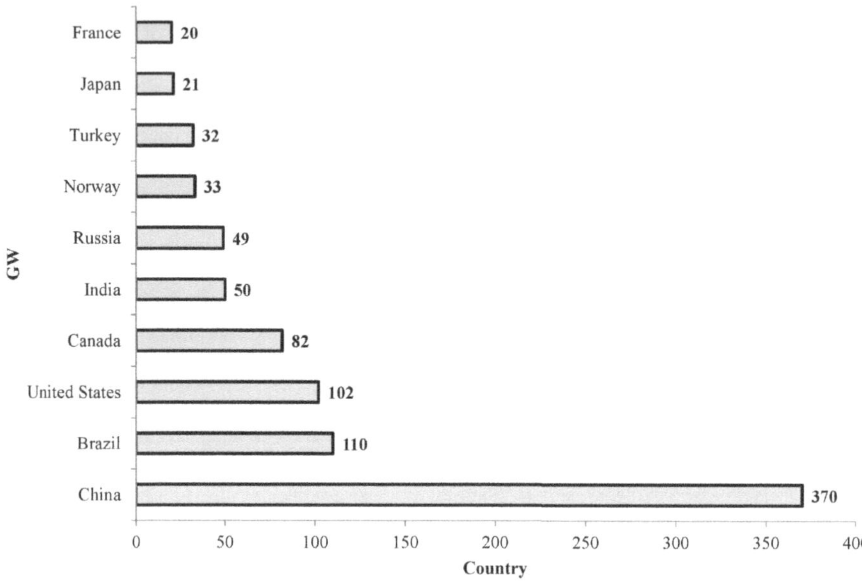

FIGURE 1.11
Total hydropower installed capacity of different countries in 2021 (Tefera & Kasiviswanathan, 2022).

is vital, especially in regions where energy demand may vary day by day. In addition, generation from hydroelectric plants can ramp up very quickly to meet sudden peaks in demand for electricity. This flexibility in the ability to supply additional capacity makes hydropower an essential complement to other renewable sources in maintaining grid stability and further reducing the use of fossil fuel resources.

1.1.6.3 Environmental and Socio-Economic Benefits

Besides their reliability, hydropower projects support a wide range of environmental and socio-economic benefits. Most hydropower projects include additional aspects in their designs for added value to their service provision for the local communities. Examples include how reservoirs created by the dam can become subjected to recreational use, such as fishing and boating, adding to the value of local tourism and economics. Hydropower also supplies water for irrigation and water supply, mainly in areas dependent on agriculture for livelihood. The regulation of water resources by hydropower projects has contributed to dependable supplies of water for irrigation, mainly in areas often hit by droughts.

Interesting Fact: Hydropower is of a multipurpose nature. One of the most interesting facts concerning hydropower is that most of it is multipurpose

FIGURE 1.12
Sardar Sarovar Dam in India.

in nature. While being applied to electricity generation, many of the projects on hydropower contribute immensely to controlling flood disasters, keeping the water supplies regulated, and supplying water for irrigation purposes. This gives even greater value to hydropower in the integrated water resource management perspective.

1.1.6.3.1 Case Study: Sardar Sarovar Dam

A classic example of a hydropower plant, apart from the supply of electricity, is the Sardar Sarovar Dam in India, as shown in Figure 1.12. This 1,450-MW dam forms part of the Narmada Valley Development Project aimed at assuring water supplies for irrigation and drinking purposes to millions of people. The controversies and litigation that have dogged its construction often relate to the submergence of village communities and ecological aspects. Simultaneously, it has also fetched kudos for promoting generation and assured regulated supplies of water. Similarly, the Sardar Sarovar Dam provides light to many households and irrigation water to Gujarat's agricultural heartland; thus, a single hydropower project could fulfil various needs of society. It is in this regard that a discussion of some of the problems affecting hydropower would be proper.

1.1.6.4 Constraints Facing Hydropower

Of course, even with all these advantages, there are also disadvantages to hydropower. Large dams fully alter the environment: large changes in the local ecosystems of the place, natural migration and movement of fishes, and even the water quality will occur. Displacing of communities may cause social tensions and conflicts. Climate change also threatens hydropower. Changes in precipitation coupled with the melting of glaciers affect the availability of water, hence consequences on the long-term reliability of hydropower generation. This situation, therefore, calls for sustainability measures and careful environmental and social impact studies with respect to hydropower projects.

Before delving into the key challenges facing renewable energy, Table 1.3 provides a comprehensive overview of the major renewable energy sources and their key performance metrics. This data helps contextualize the challenges and opportunities discussed in the following sections.

TABLE 1.3

Comparative Analysis of Major Renewable Energy Technologies: Performance Metrics, Economic Indicators, and Environmental Impact Assessment (2023–2024) (Guidi et al., 2023; Kasradze et al., 2023)

Metric	Solar	Wind	Hydropower	Biomass	Geothermal
Global installed capacity (GW, 2023)	1,250	837	1,360	143	16.9
Average cost (US$/MWh)	36–44	29–56	30–60	65–110	45–85
Typical efficiency rate (%)	15–23	35–45	90–95	25–35	10–20
Annual growth rate (%)	25–30	15–20	2–3	5–7	4–6
Carbon footprint (gCO_2/kWh)	40–48	11–14	24–28	230–250	38–45
Land use (m²/MWh/year)	40–45	70–85	10–20	450–500	15–25
Average lifespan (years)	25–30	20–25	50–100	20–25	30–50
Initial investment cost (US$/kW)	800–1,000	1,200–1,700	1,500–3,000	2,000–4,000	3,500–7,000
Operation and maintenance cost (US$/kW/year)	10–15	20–40	15–25	50–100	60–120
Capacity factor (%)	15–30	30–50	40–60	65–85	70–90
Water consumption (L/MWh)	10–20	1–2	3,000–4,500	1,900–2,200	2,000–2,500
Job creation (jobs/MW)	3.4–3.9	1.5–2.1	1.3–2.0	4.8–5.5	3.7–4.2
Storage integration cost (US$/kWh)	150–200	120–180	5–20	N/A	N/A
Grid integration cost (US$/kW)	100–150	150–200	50–100	80–120	70–100
Technology readiness level (1–9)	9	9	9	8	8
Power density (W/m²)	5–20	2–3	5	0.5–2	0–100
Market maturity index (1–5)	5	5	5	4	4
Environmental impact score (1–10)	8	9	7	6	8
Resource availability score (1–10)	9	8	6	7	5
Energy payback time (years)	1–2	0.5–1	1–2	3–5	2–3

1.1.7 Key Challenges and Future of Renewable Energy

The renewable energy sector has, in the last couple of decades, seen phenomenal growth that has, in a way, reshaped the globe's energy perspective. Yet, even with such progress, there are still various key issues to be resolved for further development and integration of renewable energy sources. For this broad-based review, we will cover three major challenges: grid integration, policy and investment, and public awareness and acceptance. Understanding these challenges can better guide us as we head towards the future of renewable energy.

1.1.7.1 Grid Integration

1.1.7.1.1 The Integration Challenge

Considering that the share of renewable sources in the world energy balance is growing, the task of their integration into existing power grids acquires a very critical dimension. Traditional power grids are designed to work with a centralized generation from fossil fuels, capable of providing a consistent output of power that can be controlled as needed. On the contrary, renewable sources of energy generate electric power through weather and daylight, which means fluctuating supplies turn grid management into a headache.

1.1.7.1.2 Technical Challenges

The integration of renewable energy requires great improvement in the grid infrastructure. The following are the technical perspectives in this regard.

1. **Intermittency**
   Solar and wind energy are not available all the time. For example, solar panels generate electricity only during sunny hours while the working of wind turbines depends upon wind speed. This intermittency develops hardships in balancing demand and supply.
2. **Voltage Regulation**
   The distributed generation from renewable resources results in voltage regulation. It is tough to maintain voltage stability if generation is scattered in the grid.
3. **Transmission Limitations**
   Geographical areas having high renewable resources are usually far from urban areas where energy demands are concentrated. Upgrading the transmission network for the interconnection of these resources with demand centres is of essence but often awfully expensive and politically disregarded.

1.1.7.2 Energy Storage Solutions

These are some of the challenges that make energy storage solutions highly indispensable. Besides, batteries and pump hydro storage can also stockpile excess energy produced during peak generation times and release it during low generation periods.

Battery storage technology in batteries is one area that has notably developed critical energy storage; a great example of this is found in South Australia – the Hornsdale Power Reserve – which boasts Tesla Batteries. This helps in stabilizing the network through instantaneous electricity supply during high demands and absorbing it during low demand.

1.1.7.2.1 Pumped Hydro Storage

This technology, invented several decades ago, relies on storing energy in the form of water pumped uphill during periods of low demand to an upward reservoir, and then released at times of peak demand. Norway and Switzerland already operate large pumped hydro systems.

1.1.8 Policy and Investment

The determinants of the progress of renewable energy depend critically on government policies. In fact, the sector shall need continued support, especially for sustained growth towards attaining a level playing field in the energy market. Indeed, policies shall be critical to successful increases in investment, innovation, and deployment of renewable technologies.

1. **Subsidies and Incentives**
   There are many forms of policies that can be used to facilitate investments in renewable sources of energy.
2. **Financial Incentives**
   Subsidies, grants, and tax credits will reduce the project cost of renewable energy. For example, the US federal government provides an ITC for solar energy systems: it gives a credit against federal taxes equal to a fixed percentage of qualified capital investment expenditures made by a taxpayer.
3. **Renewable Energy Standards**
   Many countries have been able to implement policies such as renewable portfolio standards (RPS), requiring a certain percentage of electricity to come from renewable sources. A good example is the state of California, with an RPS mandating 60% of its electricity to be qualified as renewable by 2030.
4. **Long-Term Contracts**
   Power Purchase Agreements (PPAs) ensure revenue certainly for renewable energy projects, thereby attracting private investment in the sector. Indeed, by locking in the price of the generated electricity over a set period, developers can ensure financing for the construction of projects.

Well, it is true that government support is essential; the following are the factors that impede investment in renewable sources of energy:

1. **Market Volatility**
   The fluctuating energy prices make investors averse to investing in renewable projects. For instance, during periods of low fossil fuel prices, the competitiveness of renewable energy decreases, leading to less investment.
2. **Regulatory Uncertainty**
   The changes in policies or regulations can present uncertainty for the investors. For instance, the rollback of incentives given for renewable energy in certain regions can result in a decline in project development.
3. **Access to Capital**
   Most renewable energy projects in developing countries have very limited access to finance. High interest rates and general low credit availability discourage deployment in renewable technologies.

1.1.9 Public Awareness and Acceptance

- *Importance of Public Engagement:* The acceptance of projects on renewable energy by the public is essential for the development of the same. Increased awareness of the benefits that were highlighted above on renewable energy will lead to more adoption and contribute towards the attainment of higher levels of support of projects in this line. Widespread misconceptions and worries regarding the various sources of renewable energy perhaps could change this.
- *Education Campaigns:* Educational campaigns can go a long way in fostering understanding and acceptance of renewable energy. Communication strategies can make people aware of various apprehensions and misconceptions, including but not limited to the following.
- *Misconception of Intermittency:* The use of renewable energy is very much enveloped by misconceptions that it is unreliable. Educating the masses about developments in energy storage and grid management will help in diffusing these apprehensions.
- *Environmental Impact:* There are those who would doubt the environmental effects of such renewable energy projects, like land disruption and habitat disruption. Transparency in project planning and environmental impact studies helps to build trust and prove a commitment to sustainability.
- *Community Engagement:* Community engagement during the process of planning and development has become very significant in gaining

acceptance for renewable energy projects. The following describes methods of effective community engagement.

- *Decision-Making:* Involvement of a community in decision-making levels solves problems and helps to gain support for projects. For example, taking into consideration the community of local stakeholders' oversite selection for wind farms may be better.
- *Highlighting the Local Benefits:* Showing that local communities benefit directly from renewable energy projects improves acceptance. In this case, highlighting jobs created, better revenues due to increased taxation, or community investments will be a real value addition for projects on renewables.

1.1.9.1 Overcoming Nimbyism

The sentiment of 'Not in My Backyard' sentiment can be one of the major barriers to the development of projects on renewable energy. It is because, in principle, the public may support the idea of renewable energy, but that often may falter with specific projects sitting in proximity. Through open dialogue, education, and demonstrating broader benefits to society, these concerns can be addressed.

1.1.10 Renewable Energy's Future

However, despite all these challenges put into thought, the future of renewable energy looks bright. Acting on the urge to address climate change, coupled with technology advances and falling costs, will see renewable energy continuing to play a cardinal role in the global energy transition.

1.1.10.1 Technological Innovations

The ongoing development in all these renewable technologies will continue to drive efficiency and cost. Solar panels, wind turbines, and energy storage systems will be improved; their application and performance could be enhanced. For example:

Floating Solar Panels: One of the new ways to generate solar energy without requiring large areas of land is floating solar farms. These can be located on a reservoir to reduce evaporation and, therefore, maximize overall efficiency.

Advanced Wind Turbines: Newer generations of wind turbines are being made that are more efficient and operate within a large range of environmental conditions. These include vertical-axis wind turbines and thus can be installed even in urban areas where traditional turbines cannot be deployed.

1.1.10.2 Global Cooperation

International cooperation is going to remain crucial in hastening the development and deployment cycle of renewable energy technologies. This will also contribute to the sharing of experiences and best practices for various countries regarding large-scale integration. In this context, collaboration between nations initiatives like the International Solar Alliance contributes to the promotion and acceleration of the usage and technological innovation of solar energy.

1.1.10.2.1 Investment Growth

The investment in renewable energy will continue to surge, in both public and private sectors. In implementation, many countries will also increase their ambition on climate targets, thus releasing more financial resources into the renewable sector. Financial support for a Green Climate Fund was established to assist developing countries in their efforts towards responding to climate change.

1.1.10.2.2 A Just Transition

A just transition towards renewable energy involves an intervention in social and economic disparities. For this reason, policymakers must launch initiatives for green jobs by providing support to workers in the fossil fuel industry. Training programmes and opportunities for reskilling can be provided to the workers so that they can be equipped for renewable energy sectors.

1.1.11 Conclusion

Probably the most significant change hitting the world's energy market, the development of renewable energies is driven by the pressing need to take care of climate change but also, not least importantly, to reduce dependence on finite resources that come with fossil fuels. As discussed in this chapter, these sources of renewable energy – solar, wind, and hydropower – are growing exponentially due to continuous technological advances that decrease their cost and raise environmental awareness among the public. Of course, all this does come with its few challenges: the need for innovation in how intermittency and reliability are managed within existing power grids; sustained investment; and enabling policies to make a difference to fostering a renewable energy sector. This cannot be overemphasized – the role of governments in providing incentives and creating a stable regulatory environment has been driving the increasing pace of innovation and deployment. A further important factor in community acceptance of renewable projects is their involvement through education and by open, frank communication about the benefits and impacts of these projects. Essential tools to resist obstacles are provided, and broad-based support is built. While there certainly is promise for renewable energy sources in the future, it will take many governments and industries – including communities – to overcome some of the challenges holding back that particular source from reaching its full potential. Thus, the emphasis on technology, policies, and community involvement paves a way towards a sustainable energy future. Briefly put, success in renewable energy has much more to offer than a response to environmental imperatives; it is all about new opportunities for economic growth and job creation, for a more secure energy system. Altogether, this transition to renewable energy could be a reality for many generations, rather than just another fad.

Chapter Review: Chapter 1 – Introduction to Renewable Energy

Multiple Choice Questions

Test your understanding of the key concepts from this chapter. Each question is followed by four options, only one is correct.

1. What is the primary benefit of renewable energy sources?
   A. They are always cheaper than fossil fuels.
   B. They emit fewer greenhouse gases.
   C. They require no infrastructure.
   D. They are available only in rural areas.
2. Which of the following is *not* a renewable energy source?
   A. Wind
   B. Natural gas
   C. Solar
   D. Hydropower
3. What does the phrase 'energy transition' refer to?
   A. Switching from coal to oil
   B. A shift from manual labour to machines
   C. Moving from fossil fuels to renewable energy
   D. Converting homes to natural gas
4. Solar panels convert which type of energy?
   A. Chemical to thermal
   B. Nuclear to electrical
   C. Thermal to mechanical
   D. Solar radiation to electricity

5. Which country is the world leader in wind power generation as of 2024?
   A. India
   B. Germany
   C. China
   D. Brazil
6. One of the major drawbacks of solar power is:
   A. It emits radiation.
   B. It only works with coal backup.
   C. It is weather dependent.
   D. It causes earthquakes.
7. What does the acronym 'RE' usually stand for in the energy sector?
   A. Real estate
   B. Renewable electricity
   C. Remote energy
   D. Reactive energy
8. The largest source of renewable energy used globally is:
   A. Geothermal
   B. Biomass
   C. Hydropower
   D. Hydrogen fuel cells
9. Which renewable energy source taps into the Earth's internal heat?
   A. Wind
   B. Geothermal
   C. Biomass
   D. Solar
10. Why is biomass sometimes controversial as a renewable energy source?
    A. It consumes nuclear fuel.
    B. It emits zero carbon.
    C. It can compete with food production.
    D. It depends on wind.
11. Which technology allows solar power to be stored for use at night?
    A. Coal reserves
    B. Oil generators
    C. Battery storage
    D. Gas turbines

12. One of the biggest challenges for wind energy is:
    A. Inability to build turbines
    B. Pollution
    C. Intermittency and variability
    D. Lack of global interest
13. The renewable energy source most associated with tidal forces is:
    A. Geothermal
    B. Wave energy
    C. Tidal energy
    D. Hydropower
14. A zero-emissions future depends heavily on:
    A. Using more coal efficiently
    B. Expanding renewable energy infrastructure
    C. Cutting down forests for power
    D. Banning electric vehicles
15. Renewable energy can contribute to sustainable development by:
    A. Increasing fossil fuel dependency
    B. Reducing pollution and creating jobs
    C. Supporting centralized grids only
    D. Replacing education with automation

Answer Key

Question No.	Correct Option
1	B
2	B
3	C
4	D
5	C
6	C
7	B
8	C

Question No.	Correct Option
9	B
10	C
11	C
12	C
13	C
14	B
15	B

References

Allegretti, G., Montoya, M. A., & Talamini, E. (2024). Renewable energy for a sustainable future. In *The renewable energy–water–environment nexus* (pp. 1–36). Elsevier.

Antonakakis, N., & Usikov, Y. (2024). Energy consumption, CO_2 emissions and the economic growth nexus in EU countries over the period 1995–2020. In *Handbook on energy and economic growth* (pp. 30–62). Edward Elgar Publishing.

Asmelash, E., & Gorini, R. (2021). International oil companies and the energy transition, Abu Dhabi: International Renewable Energy Agency.

Deshmukh, M. K. G., Sameeroddin, M., Abdul, D., & Sattar, M. A. (2023). Renewable energy in the 21st century: A review. *Materials Today: Proceedings, 80,* 1756–1759.

Dykes, K., Veers, P., Lantz, E., Holttinen, H., Carlson, O., Tuohy, A., Sempreviva, A. M., Clifton, A., Rodrigo, J. S., & Berry, D. (2019). *Results of IEA Wind TCP Workshop on a grand vision for wind energy technology.*

Fessler, D. C. (2019). *The energy disruption triangle: Three sectors that will change how we generate, use, and store energy.* John Wiley & Sons, Inc.

Guidi, G., Violante, A. C., & De Iuliis, S. (2023). Environmental impact of electricity generation technologies: A comparison between conventional, nuclear, and renewable technologies. *Energies, 16*(23), 7847.

Gul, M., Kotak, Y., & Muneer, T. (2016). Review on recent trend of solar photovoltaic technology. *Energy Exploration & Exploitation, 34*(4), 485–526.

GWE Council (2020). *Global offshore wind report 2020* (vol. 19, pp. 10–12). Global Wind Energy Council: Brussels, Belgium.

Hasanuzzaman, M., Zubir, U. S., Ilham, N. I., & Seng Che, H. (2017). Global electricity demand, generation, grid system, and renewable energy polices: A review. *Wiley Interdisciplinary Reviews: Energy and Environment, 6*(3), e222.

Hills, R. L. (1996). *Power from wind: A history of windmill technology.* Cambridge University Press.

Kasradze, M., Kamali Saraji, M., Streimikiene, D., & Ciegis, R. (2023). Assessing key indicators of efficient green energy production for IEA members. *Environmental Science and Pollution Research, 30*(19), 55513–55528.

Kothari, D. P., Ranjan, R., & Singal, K. (2021). *Renewable energy sources and emerging technologies.* Prentice Hall.

MacKenzie, J. J. (2021). Energy and environment in the 21st century: The challenge of change. In *Energy and environment* (pp. 15–47). Routledge.

Mahendru, M., Tiwari, A. K., Sharma, G. D., Nathaniel, S., & Gupta, M. (2024). Energy-growth nexus for 'Renewable Energy Country Attractiveness Index' countries: Evidence from new econometric methods. *Geoscience Frontiers, 15*(3), 101704.

Malanima, P. (2014). Energy in history. In *The basic environmental history* (pp. 1–29). Springer.

Mulopo, J. (2022). A mini-review of practical interventions of renewable energy for climate change in sub-Saharan Africa in the last decade (2010–2020): Implications and perspectives. *Heliyon,* 8(11), e11296.

Nwokediegwu, Z. Q. S., Ibekwe, K. I., Ilojianya, V. I., Etukudoh, E. A., & Ayorinde, O. B. (2024). Renewable energy technologies in engineering: A review of current developments and future prospects. *Engineering Science & Technology Journal, 5*(2), 367–384.

Regan, L., Wong, B., Preston, B. L., & Curtright, A. E. (2021). *Incentivizing solar.*
Sørensen, B. (1991). A history of renewable energy technology. *Energy Policy, 19*(1), 8–12.
Sorensen, B. (2004). *Renewable energy*. Elsevier.
Sutabutr, T. (2012). Alternative energy development plan: AEDP 2012–2021. *Journal of Renewable Energy and Smart Grid Technology, 7*(1), 1–10.
Tefera, W. M., & Kasiviswanathan, K. (2022). A global-scale hydropower potential assessment and feasibility evaluations. *Water Resources and Economics, 38*, 100198.
Valavanidis, A. (n.d.) *Global electricity generation from renewable sources.*
Wasti, A., Ray, P., Wi, S., Folch, C., Ubierna, M., & Karki, P. (2022). Climate change and the hydropower sector: A global review. *Wiley Interdisciplinary Reviews: Climate Change, 13*(2), e757.

2

Renewable Energy in the Construction Industry

2.1 Introduction

The design, construction, and use of structures are being impacted by contemporary building projects that use renewable energy sources, which is altering construction paradigms. Imagine a house that not only shields him and his family from the elements but also provides electricity, runs appliances, and even lowers the expense of living there. With the growing integration of geothermal heating, wind turbines, and solar photovoltaic (PV) panels into modern construction, this dream is progressively coming true. This industry, which has historically consumed a significant amount of energy and accounts for over one-third of global CO_2 emissions, is undergoing transformation because of the application of these technologies. About 39% of average global carbon emissions, including packaged and embedded emissions, come from construction activities alone (Huang et al., 2018). However, renewable energy technology can help make new building construction sustainable without largely depending on fossil fuel energy. The first step towards reducing greenhouse gas emissions is integrating renewable energy into buildings. The second step is endorsing policies and groups that are advocating for it, such as the Paris Agreement, which committed to limiting global warming to no more than 2°C. An excellent illustration of this change is the growing popularity of 'net-zero energy buildings' (Meinshausen et al., 2022). Since PV and wind systems generate as much electricity as they require, they are already self-sufficient. One example that truly defines description is the Bullitt Center in Seattle, WA, the first commercial structure to use PV panels on its roofs to produce all its own electricity. It illustrates the economics and usefulness of environment-friendly buildings. These projects demonstrate that incorporating a green design philosophy into building architecture is not only a moral conundrum but also highly profitable. The implementation of renewable systems in construction depends heavily on the technical performance, cost, and lifespan of each energy source. Table 2.1 summarizes these aspects across multiple renewable technologies.

DOI: 10.1201/9781003599272-2

TABLE 2.1

A Comparative Overview of Efficiency, Cost, and Complexity of Integration for Major Renewable Technologies Applied in Building Projects

Technology	Efficiency (%)	Capacity Factor (%)	Ideal Climate Zone	Installation Cost ($/kW)	Lifespan (Years)	O&M Cost ($/kW/year)	Land Use (m^2/kW)	Carbon Offset (kg CO_2/year)	Common Failure Modes
Solar PV	15–22	20–25	Sunny, temperate	1,000–1,500	25–30	20–30	6–8	800–1,200	Inverter failure, shading
Solar thermal	35–70	30–50	Arid, high insolation	600–1,000	20–25	15–25	4–6	900–1,400	Scale buildup, corrosion
Wind turbines	35–45	25–40	Coastal, flat	1,500–2,500	20–25	30–50	50–100	1,000–1,800	Blade fatigue, gearbox
Geothermal	45–55	70–90	All (if geology allows)	2,500–5,000	30–50	40–60	0.5–1 (borehole)	1,500–2,200	Pump wear, loop leaks
Biomass CHP	20–25	60–85	Agricultural	2,000–3,500	15–20	60–80	10–15	1,300–1,900	Feedstock inconsistency
Hybrid solar-wind	30–35	40–60	Mixed/variable	2,000–3,000	25–35	45–70	10–20	1,600–2,500	Control system faults

The issue of energy availability is also resolved by using renewable energy in buildings. Common energy sources, such as electricity, can be disastrous due to natural disasters or system failures, leaving structures open to outside disturbances. Despite this disturbance, houses with solar panels and battery storage, such as Tesla's Powerwall, continue to operate. For instance, when the larger and more complicated macro-grid failed during Hurricane Maria in Puerto Rico, solar microgrids supplied the much-needed power for essential services. They claim that installing robust renewable energy makes a building more prepared for any shocks in the future.

This chapter aims to give readers a comprehensive grasp of the various practical uses of renewable energy systems in the development of building engineering. Finally, the following points will become clearer to you:

How Renewable Energy Systems Operate in Construction: This section will describe the various renewable energy systems, such as solar, wind, and geothermal systems, and how they are used in construction projects, which can range from single homes to massive structures.

Problems and Their Solutions: Among other chapters, this one will emphasize the obstacles that define the endeavour to incorporate renewable energy, including its structure, costs, and current solutions.

2.1.1 Engineering and Design

In mid-century development, there is particular attention and concern about how to make building power systems renewable. Solar built-in facades, for instance, are solar panels that are included into the exterior framework of a building. In addition to generating electricity, each of these facades enhances the exterior of the structure. For instance, The Edge in Amsterdam has established a benchmark for green offices by constructing a net zero energy office building using a combination of solar systems. This integration requires careful design and innovative architectural solutions so that the renewable energy systems match the building's overall structure. From its contribution to global electricity generation and employment opportunities to advancements in energy efficiency and smart grid integration, the data underscores how renewable technologies are reshaping building design and operation. Notably, innovations like solar facades and AI-driven energy management are enhancing both efficiency and resilience, while policy incentives and public health benefits further strengthen the case for widespread adoption.

2.1.2 Advantages of the Environment

Carbon emissions from buildings equipped with renewable energy systems can be greatly reduced. For instance, a standard household solar system may offset about 3–4 tonnes of carbon dioxide each year, which is the same as planting more than 100 trees. Buildings that use renewable energy systems reduce carbon emissions and enhance urban air quality, both of which have a detrimental impact on public health. As cities expand and gain new insights into their effects on air quality, the use of renewable energy in construction is essential to the creation of sustainable, healthy urban regions (Baral & Guha, 2004).

2.1.3 Effects on Finances

Another advantage that is undoubtedly very useful financially is the use of renewable energy sources in buildings. Installing solar panels, for example, can lower a user's electricity expenses by up to 70%. Sometimes end-use facilities can supply excess energy to the grid, which brings in extra money for the facility. According to the IRENA analysis, for every dollar invested in renewable energy, benefits can be three to eight times the cost of the investment over the system's lifetime. These rules make it more financially viable to employ renewable energy in any type of building, whether it be residential or commercial (Weckend et al., 2016).

2.1.4 Examples from the Real World

The following example shows how renewable energy is increasingly being used in construction projects. For example, California is lowering its carbon footprint by enforcing new requirements that mandate solar systems to be included in most new home construction projects. Business buildings also use renewable energy, so it's not just for rooftops. For example, Denmark's Copenhagen Metro line is now partially powered by wind turbines, which is a perfect example of how renewable energy is being widely embraced and employed in urban environments. These illustrations show how a variety of building types of dynamics can be changed by renewable energy.

2.1.5 Thought Box

When thinking about the use of renewable energy sources in buildings, take your surroundings into consideration. Examine your house or school to determine potential locations for windmills or solar panels. What challenges could you face if you try to employ these technologies? Will you need to make any adjustments, or can you use the geometry of your roofing system to connect solar panels directly to the roof? You should be better equipped to understand some of the more pragmatic concerns related to the incorporation of renewable energy technology into building structures after completing this assignment. It is crucial that the student understands the justification for using renewable energy in construction in addition to completing the tests. Greetings to all students studying architecture, engineering, and urban planning, as we are a generation that will be shaping the design of buildings in the future. You will be ready to design buildings that can meet today's energy concerns and other sustainability issues worldwide using the knowledge you acquire from the course. Understanding renewable technology can enable you to make a positive contribution to society, whether your goal is to develop net-zero energy homes or work on exciting large-scale infrastructure projects driven by renewable energy.

2.2 Design Considerations for Integrating Renewable Energy

These are not limited to the installation of tiny renewable energy sources like solar PV, wind turbines, etc.; rather, they are a part of a much broader and more complex process of building projects. This process reimagines the design and produces a bracket solution that seamlessly integrates all building components and renewable energy systems. As with any best practice standards, bridging effectively requires a reasonable mix of architectural and specifier intervention, along with technological expertise in energy modelling applications. As a result, creative, affordable, energy-efficient, and environment-friendly construction is produced. Examples of how this integration could be achieved include the issues of architectural design, the selection of materials, and the application of energy modelling techniques in construction projects.

2.2.1 Design Recommendations for Architectural Renewable Systems

The key to incorporating renewable energy systems is architectural design. The energy systems of a building must be both visually beautiful and useful, blending in with the structure's form, volume, style, and purpose, as this text also recognizes. When integrating renewable energy, site selection or building orientation and flexibility in the location and installation method of renewable technologies are crucial considerations. When a building is orientated correctly, it can decrease heat loss and enhance natural light. To maximize the amount of heat that the sun can capture by the rooftop solar panels, for instance, a building in the Northern Hemisphere should have its roof facing south, whereas a structure in the Southern Hemisphere should have its roof facing north. Wind turbine sites are also very important. It is crucial to consider the predominant wind regime in relation to steady wind currents since coastal or high-altitude areas are usually the most promising from the standpoint of constructing wind power plants. For example, the Masdar City project in Abu Dhabi reduced energy consumption by up to 40% by optimizing solar energy and lowering cooling loads using both site and building orientation. Heat regulation, natural ventilation, and light-capturing architecture are some of the tactics used to reduce energy consumption in architectural designs. These strategies aid in improving overall energy efficiency when paired with real, functioning renewable energy installations. Examples of passive design features that don't need any form of mechanical intervention are plants, thermal mass walls, and large windows that face the sun. The Bullitt Centre in Seattle, for instance, is an energy-positive building that employs passive solar heating and roof-integrated solar electricity. The employment of cutting-edge technologies discussed above has lowered energy imports while also demonstrating that sustainability in the built environment is attainable. To achieve greater energy efficiency and sustainability in modern buildings, the integration of intelligent control systems has become indispensable. Smart building energy management systems (BEMS) now leverage artificial intelligence, IoT-enabled sensors, and real-time data analytics to optimize energy consumption across heating, cooling, lighting, and ventilation systems. These technologies enable proactive energy decision-making, load balancing, and remote management, significantly reducing operational costs and environmental impact. The architecture and components of such systems are depicted in Figure 2.1, which illustrates how data flows through various modules from sensors and user interfaces to energy control mechanisms resulting in an automated, energy-optimized building environment.

In response to public pressure to use more renewable energy, architects have started incorporating renewable energy systems into building structures. One example of a revolutionary installation technique is building-integrated

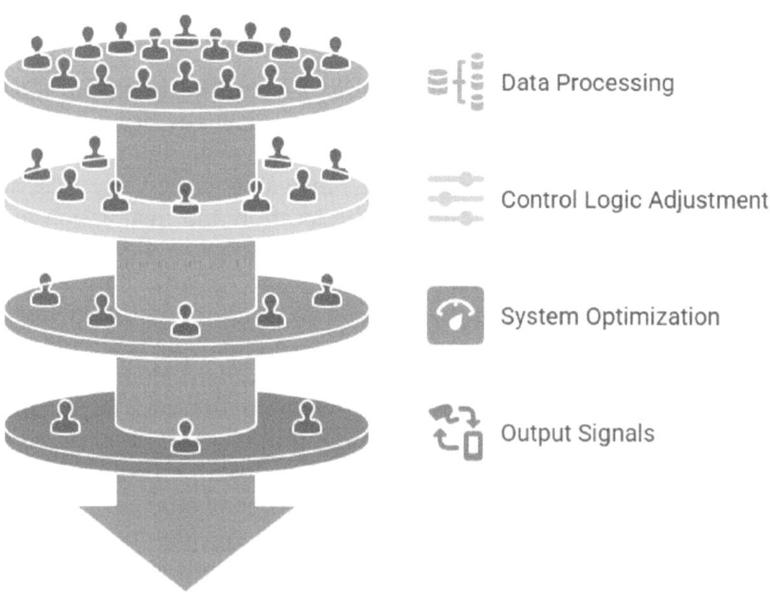

FIGURE 2.1
Smart building energy management.

photovoltaics (BIPV), which incorporates solar panels into a building's roof, facade, or skylights to function as an energy system and an aesthetic feature. An interesting illustration of BIPV in operation is the Freiburg Solar Tower in Germany. On one side of the building's outer shell, PV modules have been installed to produce power and act as the building's external skin. Furthermore, wind turbines are being incorporated inside buildings. At the Bahrain World Trade Center, for instance, three vertical-axis wind turbines (VAWTs) have been positioned between two towers, contributing roughly 15% of the building's electricity. The building envelope has an impact on a home's energy efficiency as well. When the building envelope functions well, energy losses are minimized, and existing renewable systems are maximized. New generation products, including STC systems, IG units, and hermetic barriers, are the contact elements that help create an effective barrier to energy use right away. The appropriate materials, such as for the window and roof openings, help to meet a large part of its energy needs through overall heat loss.

2.2.2 Material Selection for Renewable-Integrated Structures

The selection of materials is another critical part in using renewable energy. The type of material used in the construction of renewable energy systems and the structure determines their longevity and efficiency. Materials that can sustain their weight without producing structural distortion are necessary for renewable energy systems like solar panels and wind turbines. Commonly used materials include steel, aluminium, and similar materials that are characterized by their strength to weight ratio and reusability. For instance, mechanics plays a major role in the construction of the gigantic 17 MW solar array at Apple Park in Cupertino, which is made of incredibly lightweight materials (Moreno Sierra, 2024). In the transition to Construction 5.0, the principles of circular economy are becoming deeply embedded in project life cycles particularly when powered by renewable energy. Unlike the linear consumption model, circular construction emphasizes reuse, recycling, and material efficiency. When integrated with renewable energy strategies, this approach fosters a regenerative construction loop that minimizes waste and carbon emissions while enhancing resource productivity. Figure 2.2 captures this synergy, demonstrating how renewable energy inputs (such as solar or wind) can align with circular workflows in design, procurement, and operations to create a closed-loop, low-impact construction ecosystem.

Just as important as the structural features of buildings is the sustainability of the materials that will be used. Building with low-switched embedding materials is one method to improve the sustainability of constructions that integrate renewable energy sources. These materials enable material reuse, hence protecting the environment from some material misuse. The eco-friendly building material known as cross-laminated timber, or CLT, is just getting started. For example, Canada's Brock Commons Tallwood House

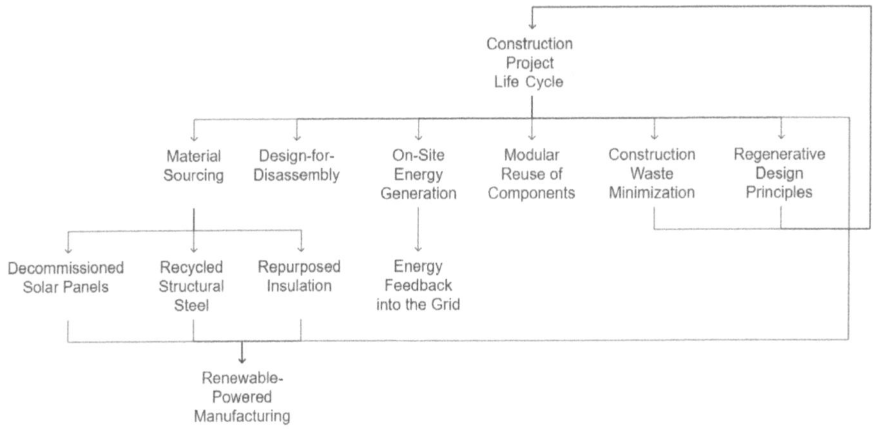

FIGURE 2.2
Circular economy in renewable energy-driven construction.

uses CLT and rooftop solar electricity–producing systems, proving that the materials are sensible alternatives for a sustainable progressive height building. Advances in solar power conversion technologies may make it easier to incorporate PV technology into buildings. For instance, a new generation of transparent solar panels is available that may be used as windows to let light into the building in addition to producing electricity. Likewise, the addition of lightweight, portable perovskite solar cells to building facades enables the integration of solar energy into urban areas. The Netherlands Pavilion at Expo 2020 in Dubai, where the panels were designed to function as both a power source and an aesthetic component of the pavilion's construction, is a great example of the several uses of TSP. Two further characteristics that seem to need serious research are insulation and heat conduction as discussed in Table 2.2. For example, thermal storage materials such as concrete or stone can efficiently support renewable heating systems by absorbing heat during the day and releasing it at night; aerogels are lightweight and have excellent

TABLE 2.2
Emission Reductions per Building by Renewable Technology and Conversion Method

Technology	CO_2 Saved (tonnes/year)	CH_4 (kg/year)	NO_x (kg/year)	Trees Equivalent	Emission Reduction (%)
Solar PV	15	9	2.5	700	40%
Wind (micro)	12	8	2.2	550	32%
Geothermal	20	5	1.2	900	55%
Biomass CHP	18	4	1.0	850	50%

insulating properties. These materials not only encourage the use of renewable power sources but also aid in reducing the overall energy consumption of buildings (Guamán-Rivera et al., 2022).

THOUGHT BOX

CIRCULAR ECONOMY: GREEN OR JUST TRENDY?

Circular design is often praised, but how do we measure real sustainability outcomes versus aesthetic or marketing-based 'greenwashing'?

2.2.3 Construction-Related Energy Modelling and Simulation Tools

The process of energy modelling is crucial for optimizing the performance of renewable energy systems in buildings. Energy modelling can predict a building's energy efficiency when it is still in the blueprint stage of design. This aids architects and engineers in selecting the most cost-effective and energy-efficient options. The evaluation technique called 'energy modelling' involves modelling a structure to analyse how much energy it uses. This enables the various renewable energy systems to be validated, their chances of success evaluated, and their future performance under practical conditions projected. The building design optimization exercise makes use of the power simulations listed below. For instance, EnergyPlus is a comprehensive building modelling programme that incorporates heating, cooling, and renewable energy system features. Energy modelling is integrated with daylighting and carbon footprint studies to provide a synchronized view of the building's environmental impact. When evaluating the feasibility of renewable energy systems, RETScreen concentrates on the financial aspect of the situation. The wind flow, which employs computational fluid dynamics (CFD) to identify the optimal site for wind turbine installation, is an additional application tool. Applications for daylighting analysis include solar plastering, wind power adaptation, and geothermal applications. Power and shading anonymity simulations rank the locations for solar panel placement, while turbine simulations assess areas with the optimum wind potential for the turbines. The estimation of geothermal systems' capacity for heat and cooling is another benefit of modelling for fine system design. The Beitou Library in Taiwan, for instance, reduced its energy usage by 30% by including solar panels, daylighting, and ground source cooling through the use of energy modelling.

Energy modelling has a lot of advantages. As a scheduling tool, it may also be used to find inefficiencies or improper operations during the construction and operation stages and make sure that renewable energy systems are operating as efficiently as possible while wasting the least amount of energy and materials. Similarly, energy modelling assists building projects in obtaining sustainable certifications such as LEED or BREEAM to guarantee that the structure satisfies the most stringent standards for sustainable energy consumption and energy efficiency. Since buildings account for 28% of global CO_2 emissions, they should prioritize renewable energy sources (Maglad et al., 2023). The building industry may create low-energy, more sustainable, and distinctive structures if architecture, materials, and energy modelling are carefully considered and put into practice. This serves as the foundation for a new generation of sustainable development construction as well as a reaction to the problems caused by climate change. Think about how you would include geothermal systems, wind turbines, solar panels, and other renewable energy sources into your design. The journey towards intelligent energy management in buildings has not been sudden but rather a result of decades-long innovation. From manually controlled thermostats to AI-driven building automation, this evolution reflects both technological progress and a growing awareness of environmental responsibility. As systems have become more adaptive and decentralized, they now offer deeper integration with renewable sources and user preferences. This historical transformation is charted in Figure 2.3, which traces key milestones in building energy management over a 30-year timeline, providing insights into how the sector has matured in its response to efficiency and sustainability imperatives.

THOUGHT BOX

HUMAN SKILL VERSUS AUTOMATION IN ENERGY MANAGEMENT

As smart systems automate decision-making in buildings, what new skills should human operators develop to stay relevant in the Construction 5.0 workforce?

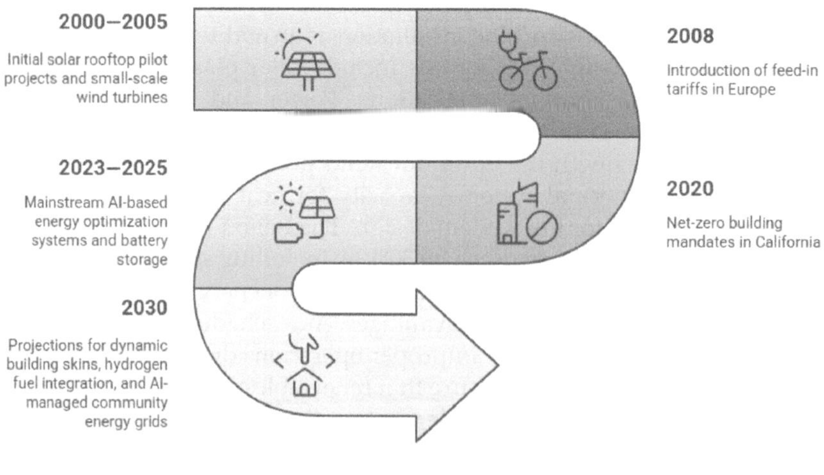

FIGURE 2.3
Thirty-year evolution of building energy management.

2.3 Renewable Energy Applications in Construction

Renewable energy systems have been adopted by construction projects as a means of implementing sustainable practices. Solar, wind, and geothermal energy are the three main forms of renewable energy sources. Each has pros and cons that are frequently used in contemporary architectural designs. When architects and engineers construct buildings to satisfy the goals of sustainable development, such technologies can be simply integrated to create environment-sensitive and energy-efficient structures. Using a few case studies, this part will explain the systems, design components, and thorough integration of solar, wind, and geothermal energy in buildings. Different regions implement diverse policy incentives and mandates to promote renewable energy adoption. Table 2.3 compares major economic regions based on financial mechanisms, net metering availability, and ease of compliance.

2.3.1 Solar Power in Structures

Solar energy, which has several uses, is one of the most popular renewable resources in the construction industry. PV systems are installed on building facades to produce power, lower demand for it, and improve sustainability. The two primary ways to use solar energy in buildings are through rooftop and building-mounted solar systems. Installing PV panels on building rooftops is Solairedirecte's most popular service. Rooftop solar power, which converts light into electricity using PVs mounted on the roof, is preferred by homeowners and business owners. Such systems are more advantageous

TABLE 2.3

The Policy Landscape for Renewable Integration in Buildings across Global Regions

Region	Tax Credit (%)	Feed-In Tariff ($/kWh)	Green Building Mandate	Net Metering	Incentives for Storage	Subsidy per kW Installed ($)	Carbon Tax ($/tonne)	Compliance Difficulty
North America	26	0.12	Partial (City/state)	Yes	Yes	300–600	50	Medium
Western Europe	30–35	0.15–0.18	Yes (EU-wide)	Yes	Yes	500–900	70	Low
East Asia	20	0.08–0.10	Partial (Japan, S. Korea)	Varies	Limited	250–500	40–60	High
Middle East	10	None	Rare	No	No	100–200	None	Very high
Sub-Saharan Africa	5–10	0.05–0.08	None	No	No	50–150	None	High

when the roof area is spacious, open, and directly facing the direction of the sun's maximum slope. In countries in the Northern Hemisphere, for example, rooftops facing south are the most effective way to capture solar energy (Ansari et al., 2013). As energy demands become more complex in modern buildings, traditional control systems are being replaced by holistic energy management funnels that account for multiple variables simultaneously. These include occupancy behaviour, weather forecasts, energy tariffs, and health system indicators. This systems-based approach allows for greater scalability and adaptability in both residential and commercial settings. Figure 2.4 presents a smart energy management system funnel that visually breaks down the layered input and decision-making processes. The funnel narrows towards optimized outputs, such as reduced peak loads and emissions, highlighting the efficiency of multilayered control systems.

BIPV systems are more advanced than conventional solar integration, which entails incorporating solar cells into building structures. BIPV systems incorporate windows, roofs, and facades in addition to performing the roles of power generation systems and structures. For instance, because BIPV panels are both functional and aesthetically pleasing, the uncontested Solar Decathlon competition has forward-thinking homes that incorporate them into their designs (Biyik et al., 2017). The Solar Tower in Freiburg, Germany, is another breathtaking example of a structure that uses BIPVs outside for both aesthetic and electrical purposes. The usage of solar energy in buildings has grown dramatically as solar panels have advanced to the point of

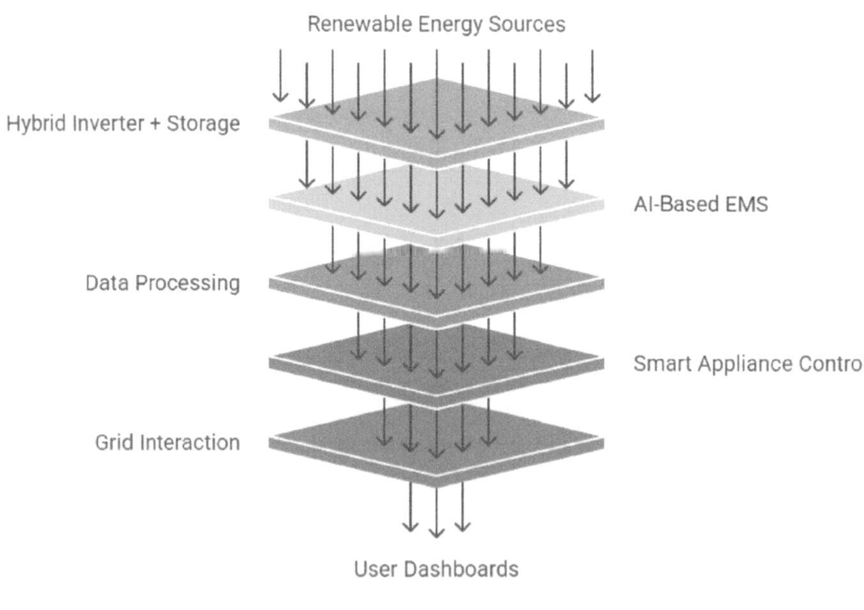

FIGURE 2.4
Smart energy management system funnel.

total transparency. Indeed, these panels can be used in place of the usual glass found in windows and skylights, letting in natural light while generating electricity at the same time. *Conclusion:* Transparent solar panels are particularly well-suited for usage in urban settings because the installation of PV composite structures in high-rise buildings frequently faces certain difficulties because of spatial concerns. Solar technology can be utilized not just to generate electricity but also for shading, which lowers heat transfer and improves a building's efficiency (Debbarma et al., 2017). Popular sunshade constructions like louvers, overhangs, and canopies can also be powered by PV cells. Solar shading is the process of modulating light and power generation via the use of movable fixtures that are mounted on windows or their facade. For instance, solar louvers are incorporated into the building of One Angel Square in Manchester, significantly lowering energy use. PV-cell-equipped walkways and carports are used by commercial structures, such as shopping malls, to provide covered areas for customers and to power the building. Conversely, dynamic shading systems are the most recent developments in solar technology. These systems use an angle that a sensor can detect to determine the amount of light that can be produced and the comfort level inside the structure at any given time. This allows the shading devices to be suitably aligned dependent on the location of the sun. One of the best examples of this concept is the Al Bahr Towers in Abu Dhabi. Their solar shading system has revolving brises that cut cooling needs in half while also producing energy from the sun.

THOUGHT BOX

IS NET-ZERO ENERGY TRULY ACHIEVABLE?

With increasing material costs and technical challenges in energy storage, can every building realistically aim for net-zero energy? Consider economic feasibility across different regions.

2.3.2 Wind Energy in Construction Projects

The fact that wind energy, which most people think of when they think of the construction of large wind farms, is gradually finding its way into construction projects is even more surprising. Since wind may be utilized to generate electricity, buildings with wind turbines are advised in areas with consistent wind. Even in situations where large wind farms are not feasible due to space limitations in urban settings, people can nevertheless benefit from well-integrated wind energy microsystems.

2.3.2.1 Vertical- and Horizontal-Axis Wind Turbines for Urban Areas

A VAWT and a horizontal-axis wind turbine (HAWT) are the two main types of wind turbines used in construction. Despite having different development processes, both kinds have uses and benefits in urban environments. What features of Voith turbines make them especially suitable for urban environments? VAWTs are tiny and able to run in any wind direction, in contrast to other types of wind turbines. VAWTs can operate effectively in any wind direction, unlike horizontal axis turbines, which makes them appropriate for use in areas with unpredictable wind patterns. In London's Strata SE1, for instance, VAWTs are situated atop a building that produces roughly 50 MWh of electricity a year (Kumar et al., 2018).

HAWTs, the second type of wind turbines used to produce power, are favoured in areas with regular wind patterns, such as coastal and suburban areas. The Pearl River Tower in Guangzhou, China, draws air into its horizontal rotors through vents in wind tunnels. These turbines assist the building in meeting its energy requirements because of their high-power output. Architectural design influences the effectiveness of wind energy systems, a form of renewable energy technology, in buildings. Architects and engineers use a range of techniques to improve wind flow around turbines and optimize energy collection. In many cases, wind is directed into turbines and drag is decreased by using aerodynamically friendly building designs (Xu et al., 2021). Buildings that are tapered or curved, like the Bahrain World Trade Center, are either aesthetically pleasing or increase efficiency by lowering air drag. Three large wind turbines between the Bahrain World Trade Center's two towers provide 15% of the structure's electricity. Turbine positioning is yet another crucial element because of the turbines' strategic arrangement. Turbines are frequently placed on roofs, in the area between two buildings, or in wide open areas with high wind speeds. It should be noted that accessibility affects wind energy systems; urban design can help optimize accessibility. For example, buildings and streets are designed to provide wind corridors, which give rooftop wind turbines a strong airflow to boost their output (Xu et al., 2021).

2.3.3 Geothermal Energy in Construction

Geothermal energy is a practical method of heating and cooling buildings because the earth's temperature is constant, and PV is a good technique to produce electricity. Geothermal systems are very good at controlling indoor temperatures all year round and using less energy. In the context of sustainable building systems, geothermal energy has emerged as a highly efficient solution for heating and cooling. By tapping into the earth's constant subsurface temperatures, geothermal heat pumps can significantly reduce energy consumption. These systems operate through a network of ground loops and heat exchangers that transfer energy between the ground and the building environment (Stober & Bucher, 2013). Figure 2.5 provides a technical schematic of a geothermal heat pump system, showing how the fluid loop circulates energy through underground pipes and interacts with interior HVAC

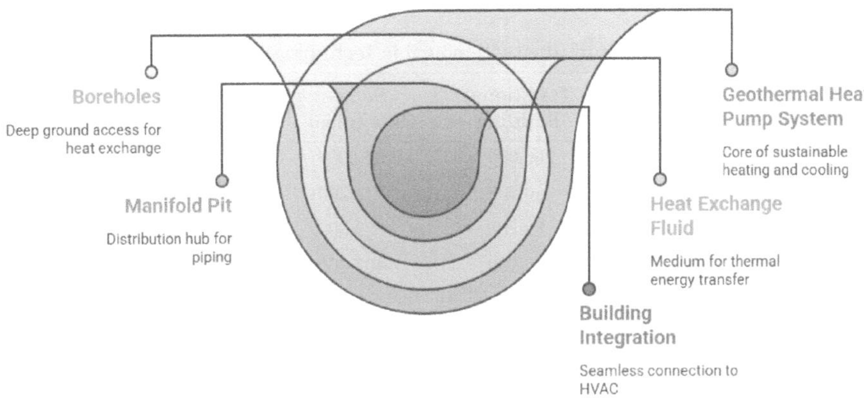

FIGURE 2.5
Geothermal heat pump system.

systems. This technology exemplifies how natural systems can be harnessed to support sustainable building operations.

THOUGHT BOX

RETHINKING LIFE CYCLE ASSESSMENTS

Are current life cycle assessment tools truly reflecting the environmental impact of renewable energy integration, or do they need recalibration for emerging technologies?

2.3.3.1 Buildings with Geothermal Heating and Cooling Systems

There are two types of geothermal heat pumps: directly excited ground heat pumps and ground-source heat pumps. The latter are by far the most common in buildings. In these systems, heat is drawn from the earth and transferred to the building in the winter and vice versa in the summer via circulating fluid pipelines. GSHPs are incredibly efficient; they can use up to four units of power to provide four units of heating or cooling. Cornell University is a great example of GSHP integration; there, energy costs have been cut by 40% and many buildings are heated and cooled down by geothermal systems (Galantino et al., 2021). The Reykjavik District Heating System in Iceland, which supplies geothermal energy to thousands of buildings, demonstrates the scalability of this technology, as discussed in Table 2.4. Geothermal energy can be used for single-building applications as well as district region heating and cooling, which includes some communities or colleges. In locations with hot springs, such as Iceland and New Zealand, these systems are very helpful (Sarbu & Sebarchievici, 2014).

TABLE 2.4
Emission Reductions per Building by Renewable Technology and Conversion Method

Category	Traditional Building	Solar-Integrated	Geothermal HVAC	Net-Zero Building
Initial capital cost ($)	260,000	320,000	360,000	420,000
Annual energy cost ($)	6,000	3,200	2,900	1,200
O&M cost/year ($)	3,200	3,800	4,000	4,500
Net present cost (30 years)	445,000	403,000	395,000	388,000
Energy cost reduction (%)	–	47%	51%	80%
ROI period (years)	–	8	10	12

2.3.3.2 Integration of Geothermal Design of Structures

Geothermal designs are often incorporated into structural plans since geothermal installations are conducted underground due to the necessity of installing distribution and circulation loops underneath. Due to their compact size and ability to be installed beneath basements and parking lots, vertically well models are a common type of application. The positioning of boreholes, the feasibility of HVAC structures, and maintenance-related factors all are part of structural geothermal designs (Gerber & Maréchal, 2012). For example, during peak hours, the University of Nottingham uses the thermal energy stored in boreholes beneath its buildings. The use of hybrid systems, which integrate geothermal energy with other renewable resources like solar or wind, is becoming more and more common. Since this technology leverages the benefits of several energy sources, it delivers significant advances in terms of dependability and energy efficiency (Aresti et al., 2024). The EcoGrid Project in Denmark, which combines geothermal energy with solar and wind turbines to create a completely renewable energy infrastructure, is likely one of the most impressive instances. To better understand the effectiveness of renewable systems across building types, Table 2.5 compiles penetration levels, emissions reduction, and payback periods for different use cases.

2.4 Case Studies and Practical Examples

Most of the world's infrastructure, business, and residential projects employ the renewable energy system as an application rather than a concept. Here, case studies and examples will shed light on the use of sustainable technology in smart homes, transit hubs powered by renewable energy, hospitals, and even green office buildings. The ramifications of renewable energy on construction are covered in this chapter, along with how this innovation

TABLE 2.5

Renewable Energy Technologies Impact Energy Use and Emissions across Various Building Typologies, Indicating System ROI and Retrofit Feasibility

Building Type	Average Renewable Share (%)	Energy Use Reduction	Technology Stack Used	Building Size Range (m^2)	Retrofit Cost per m^2 ($)	Payback Time (Years)	Average Emissions Reduction (%)
Single-family	45	30–50%	Rooftop PV, solar thermal	100–250	150–300	6–8	35–50%
High-rise	60	50–65%	PV + geothermal + wind	2,000–10,000	200–450	10–12	55–70%
Industrial Park	35	20–40%	Biomass CHP, wind	10,000+	300–600	12–15	40–60%
Mixed-use urban	55	40–55%	PV + storage + geothermal	1,000–5,000	250–400	9–11	50–65%

stimulates new projects by serving as a constant reminder to people through real-time applications.

2.4.1 Residential Construction Projects

2.4.1.1 Smart Homes with Renewable Energy Integration

The smart home is the next generation of residential structures, combining modern technology and renewable energy sources to boost convenience and encourage sustainability. These residences combine intelligent technology that maximizes energy use with renewable resources like solar and wind power. In California, household solar integration has been transformed by Tesla Solar Roofs. In addition to providing an additional source of alternative power generation within a home, Tesla solar tiles are affixed to a home's roof and have a sleek, understated appearance. By connecting to this tile, the Powerwall battery system creates an extra reserve that can be used at night or even during blackouts (Ali et al., 2021).

Through monitoring, one can get real-time production and consumption data, revealing rich aspects within the organization. For example, the Ecovillage in Ithaca, NY, is a neighbourhood that uses energy-efficient building techniques, solar panel systems, and geothermal heating to create sustainable homes. Rooftop solar panels and geothermal loops for heating and cooling provide most of the community's electricity. To save energy, smart homes include automated systems that turn on and off lighting, heating, and other appliances. The shift towards renewable energy in the construction sector is accelerating, driven by both policy imperatives and environmental awareness. Solar panels, wind turbines, biomass generators, and hydroelectric micro-systems are increasingly being integrated into both new and retrofitted structures. This transition not only decarbonizes building operations but also empowers localized energy generation. Figure 2.6 showcases a comparative overview of renewable energy adoption trends in construction, highlighting various energy sources, their adoption rates,

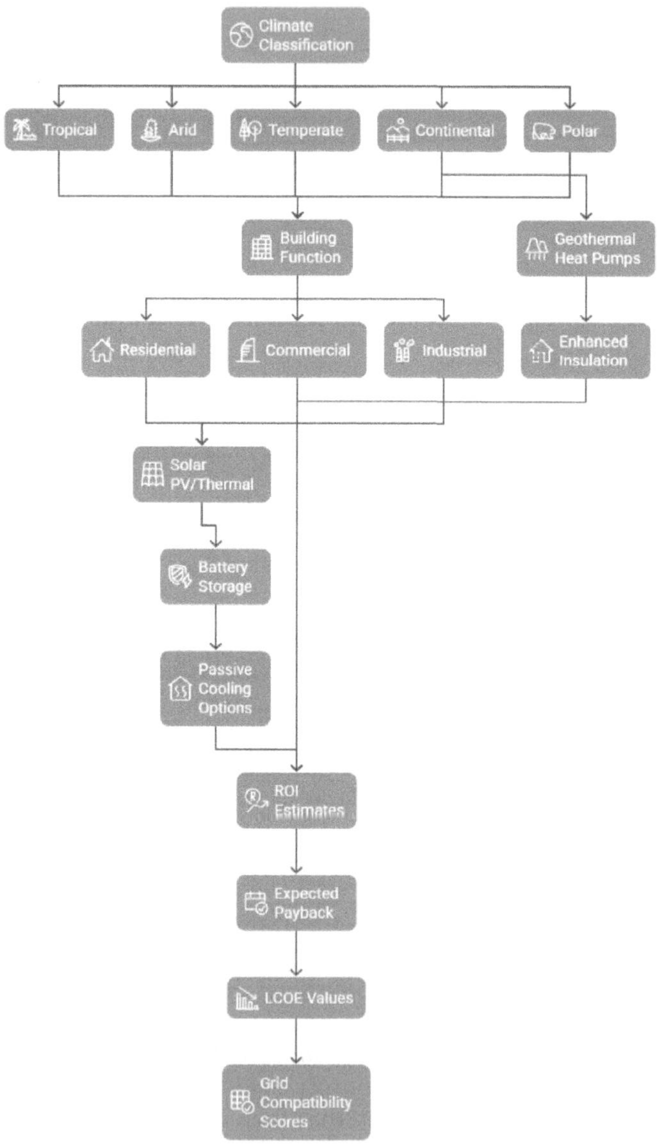

FIGURE 2.6
Renewable energy adoption in construction.

and integration strategies. The visual underscores the growing momentum towards off-grid and net-zero energy buildings.

2.4.2 Commercial and Industrial Construction

2.4.2.1 Green Office Buildings

The goal of green office buildings is to give workers a productive and healthy work environment with as little negative environmental impact as possible. By using renewable energy sources, energy-efficient design, and recycled materials, these buildings are also energy-recycled. Amsterdam's 'the Edge' office skyscraper, which is regarded as the world's greenest, has more than 65,000 square feet of solar roof spaced over its side to generate more energy than it uses. Additionally, The Edge has geothermal heating and cooling. Therefore, regardless of the season, its surrounds have always been kept up, with its energy, light, and temperatures well managed. Renewable energy technology is incorporated into the architecture of the Bank of America Tower in New York City. The cogeneration facility uses renewable biofuels to generate power and heat. Additional amenities include a rooftop rainwater collection system and energy-efficient LED lighting both inside and outside. This building received a LEED Platinum rating after meeting stringent sustainability standards (Eichholtz et al., 2010).

2.4.2.2 Renewable Systems in Warehouses and Factories

Industrial structures, such as factories and warehouses, are among the structures that require a lot of energy. Such structures can significantly reduce their manufacturing costs and the environmental effect of their construction by implementing a renewable system. The Tesco Grocery Distribution Centre in the UK is among the best examples of renewable energy in a warehouse. The large solar array at the middle provides most of the energy requirements. Furthermore, the facility gets electricity from a large wind turbine that is built on the grounds. Insulation and energy-efficient lighting reduce energy use even more. Factories are also converting to sustainable energy sources. The Frito-Lay Factory in Casa Grande, AZ, is powered by solar energy and a biogas recovery system. These solar panels provide electricity while the biogas system converts methane from waste products into energy to operate the facility. As a result, the facility has achieved an 80% reduction in energy consumption, setting a high bar for ecologically responsible corporate operations (Tang & Zhou, 2012). Building practices have steadily evolved to prioritize sustainability, moving from passive design strategies to data-driven performance optimization. Green certifications, life cycle assessments, and environmental impact analyses have become standard metrics in this evolution. This historical and conceptual progression is captured in Figure 2.7, which presents the evolution of sustainable building practices through

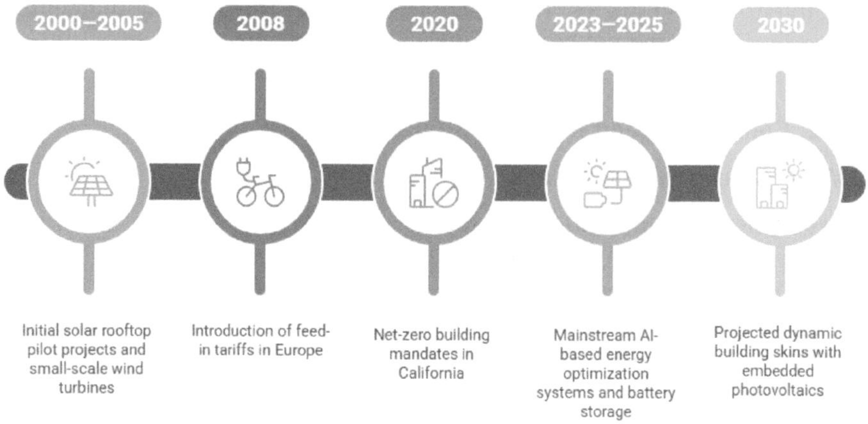

FIGURE 2.7
Evolution of sustainable building practices.

successive stages from early conservation techniques to advanced smart systems. The diagram illustrates how each phase incorporates more sophisticated tools and frameworks, culminating in the current era of responsive, intelligent buildings aligned with sustainability goals.

2.4.3 Infrastructure and Public Projects

2.4.3.1 Energy-Efficient Public Infrastructure

Renewable energy technologies, which reduce running costs and have a reduced environmental impact, are increasingly being incorporated into the design of public infrastructure projects such as schools, hospitals, and community centres. The facade of the International School of Denmark in Copenhagen is covered in solar panels; up to 50% of its needs are produced locally to lessen its carbon footprint (Nørgaard & Poder, 2019). The panels serve as a teaching tool for the kids, who are studying solar energy as part of their curriculum. Another important issue is the utilization of renewable energy sources in hospitals. Butaro Hospital in Rwanda uses solar panels to power essential medical equipment and lighting. It also reduces the need for air conditioning using passive cooling methods. A vital component of rural healthcare, Butaro Hospital's use of renewable energy ensures service delivery even in the case of a power outage. Another excellent example of public infrastructure powered by renewable energy is the Sweden Växjö Concert Hall and Library. The building's geothermal heating and cooling systems and rooftop solar panels provide most of its energy. This multiuse structure is an illustration of how renewable energy sources can be extended to create large public spaces. The integration of renewable energy systems in buildings is no longer a matter of isolated installations, but rather a coordinated

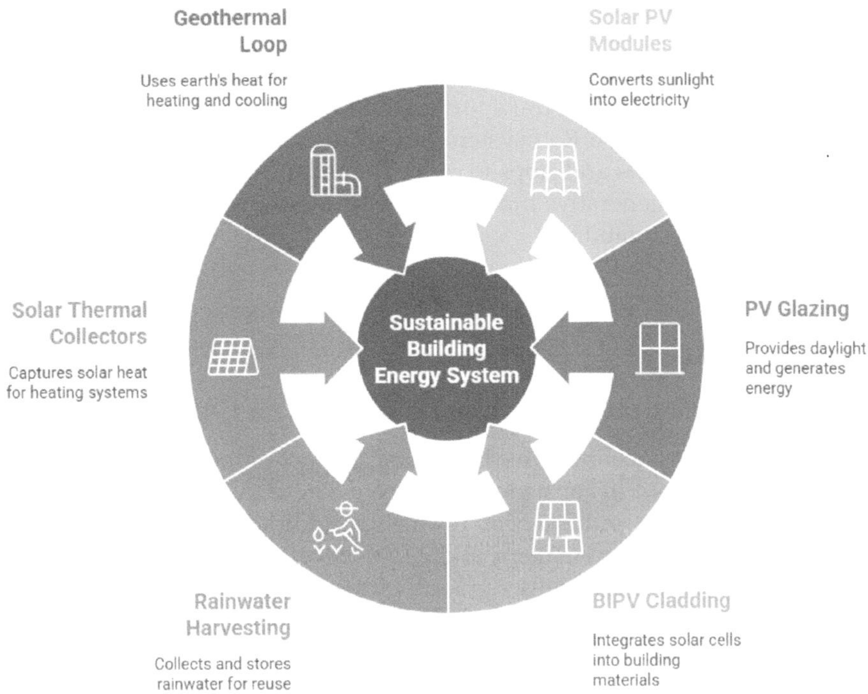

FIGURE 2.8
Integrated renewable energy systems in buildings.

network of generation, storage, and distribution technologies. Solar PV panels, wind turbines, battery systems, and smart grids now operate in tandem to ensure efficient, resilient, and autonomous energy flows. Figure 2.8 details such an integrated system architecture, where different renewable components are synchronized with building management systems to optimize load sharing, reduce dependency on fossil fuels, and enhance energy security. This approach reflects the systemic thinking necessary for Construction 5.0.

2.4.3.2 Transportation Hubs Powered by Renewable Energy

Transportation hubs such as airports and train stations consume the most energy in urban regions. Incorporating renewable energy systems into such structures has two main advantages: it reduces the impact on the environment and improves operating efficiency. Cochin International Airport in India is the first airport in the world to operate exclusively on solar power. The solar array, which includes over 46,000 panels, is net-positive, meaning it generates more energy than it consumes. This project has served as an inspiration for other airports throughout the world, including Denver International Airport in the United States. The solar roof atop London's Blackfriars Railway Station, which has more than 4,400 PV panels, can generate 1.1 MW of electricity per year, or half of the station's total electricity use (Butt et al., 2024). This would meet the needs of station clients while partially shading them with renewable energy sources. Public transport systems also employ renewable energy. The Los Angeles Metro Rail system has solar panels installed along all its maintenance facilities and stops to reduce its reliance on grid electricity. The metro also has regenerative braking devices to capture train energy to further improve sustainability.

2.5 Environmental and Economic Impacts on Construction

Renewable energy technologies have a big impact on economics and environment when used in construction projects. These include long-term resilience, sustainability, and financial rewards in addition to immediate energy savings. Developers, lawmakers, and end users all prefer renewable energy in construction because of its reduced carbon footprints, enhanced urban sustainability, and strong financial benefits. This section looks at how renewable energy systems in buildings affect the economy and the environment.

2.5.1 Advantages for the Environment

The construction business is the one that contributes the most to greenhouse gas emissions of any industry, accounting for over 39% of global CO_2 emissions (including operational and embodied emissions). Thus, buildings can help achieve global sustainability objectives and climate commitments by reducing their environmental impact with the use of renewable energy sources (Ibn-Mohammed et al., 2013).

2.5.1.1 Reducing Carbon Footprints

Renewable energy solutions are the cornerstone of any plan to lower a building's carbon footprint. Building operational emissions could be greatly reduced by replacing fossil-fuel-based sources with clean energy. For instance, it is estimated that a typical home solar panel system can offset about 3–4 tonnes of CO_2 per year, which is equivalent to growing more than 100 trees per year (Baral & Guha, 2004). There are currently several examples of carbon-neutral or even carbon-negative green buildings that have combined solar, wind, or geothermal energy systems. The Bullitt Center in Seattle, one of the greenest commercial buildings in the world, with a net reduction in carbon emissions overall and a positive net generation balance. Amsterdam's Edge Building, which was able to cut its operating emissions by more than 50% by combining energy-efficient technologies with renewable systems, provides more proof that renewable energy is influencing urban construction. The environment also gains from the conservation of non-renewable resources. Renewable energy sources, such as solar panels and wind turbines, prevent the depletion of finite energy supplies because they consume infinite natural resources. This shift reduces carbon emissions and environmental harm caused by the extraction and processing of fossil fuels. The enhancement of urban sustainability enables broader urban sustainability by promoting cleaner air, reducing heat island effects, and improving resource efficiency. Urban areas have higher energy demands and pollution levels than other locations due to their dense infrastructure and huge population. Building integrated renewable energy systems can tackle these issues by producing clean energy and lowering total energy use. Copenhagen and Vancouver are two examples of cities that have effectively incorporated renewable energy to advance urban sustainability. Over the past few decades, they have significantly reduced citywide carbon emissions by a staggering 30% by including renewable energy sources like solar panels, wind turbines, and district heating systems into their urban development strategies. Incorporating renewable systems into urban environments also contributes to the improvement of biodiversity and natural ecosystems. For instance, by providing insulation, reducing heat absorption, and creating spaces for plants and animals, solar-powered green roofs enhance ecological equilibrium. These solutions combine energy efficiency and environmental preservation to help build liveable, sustainable societies. Material sustainability is not only influenced by origin but also by the energy used in processing and installation. Table 2.6 compares life cycle carbon emissions of common materials, with and without renewable-powered processing.

2.5.2 Financial Aspects

The financial benefits of employing renewable energy in construction are just as compelling as the numerous recognized environmental benefits. In this way, because renewable systems provide substantial cost savings over their lifespan, developers and property owners have strong financial incentives to implement them.

2.5.2.1 Cost-Benefit Evaluation of Renewable Energy Systems in Building

Concerns over affordability are frequently raised, even though many of these projects involving the installation of renewable energy systems are frequently very costly. Comprehensive cost-benefit analysis, however, demonstrates that these technologies eventually yield large returns on investment. According to IRENA, solar PV systems are now among the least expensive energy generation solutions, with an average cost reduction of over 80% in just the last ten years (Dincer, 2011). For example, household solar panels pay for themselves in six to eight years, thanks to energy savings and government

TABLE 2.6

Embodied Carbon for Key Construction Materials

Material Type	Standard Energy Source	Life Cycle Emissions (kg CO_2e/kg)	With Renewable Energy (kg CO_2e/kg)	Emission Reduction (%)	Main Emission Phases
Concrete (Standard)	Coal-fired electricity	0.95	0.61	36%	Cement production, curing process
Reinforced steel	Fossil-fuel electric arc	1.85	1.08	41.6%	Smelting, rolling
Aluminium (primary)	Fossil electricity mix	11.50	6.30	45.2%	Electrolysis, refining
Engineered timber	Diesel-powered processing	0.47	0.30	36.2%	Drying, gluing
Glass (architectural)	Natural gas + grid mix	1.60	1.00	37.5%	Melting, annealing
Bricks (clay-fired)	Natural gas	0.49	0.33	32.6%	Kiln firing
Insulation (polyurethane)	Fossil-derived heat	3.20	2.25	29.7%	Blowing agents, polymerization

incentives like tax credits and feed-in tariffs. For the remaining years of the system's usable life, which might be more than 25 years, electricity remains free. Large-scale commercial or industrial enterprises can also benefit from renewable solutions. The Tesla Gigafactory in Nevada, for instance, uses exclusively solar and wind energy, which drastically lowers operating costs. Due to its on-site generating, it does not purchase energy and experiences less price volatility for power. These benefits ensure the factory's long-term profitability and, thus, the viability of renewable technology, even for very large corporations. More importantly, the addition of renewable energy sources increases real estate values. According to multiple surveys, properties with solar panels, for example, sell for up to 4% more than ones without, reflecting the increased demand for energy-efficient real estate (Wilkinson & Sayce, 2020). For commercial and public projects, renewable energy certification through LEED or BREEAM, among other similar programmes, increases rental and occupancy rates, increasing the financial returns. Reduced Life Cycle Expenses Renewable systems offer substantial savings during and after their operational lives by reducing energy consumption over the duration of their long maintenance life, which is financially advantageous to building owners. Unquestionably, one of the primary advantages of renewable systems is their extremely low running costs. Solar panels need extremely minimal maintenance and don't require fuel. Over the course of their lifetime, solar systems usually result in a 70–90% reduction in electricity expenses, depending on the area and system size (Lai et al., 2017). Wind turbines and geothermal systems are also quite economical when compared to fossil-fuel-based heating or cooling systems. The geothermal-powered Reykjavik District Heating System in Iceland is a prime example; it has continuously reduced heating expenses for the residents by over 40% when compared to conventional systems. Continuous financial gains are guaranteed by a geothermal system's long operational life, which frequently surpasses 50 years. It also includes life cycle benefits, such as avoided costs due to environmental compliance and the carbon price. As governments throughout the world implement stronger measures to limit emissions, buildings that use renewable energy sources stand to benefit and avoid fines. Energy modelling supports effective decision-making in renewable integration. Table 2.7 provides a comparison of popular modelling tools based on capability, compatibility, and renewable simulation features.

2.6 Challenges and Limitations

Renewable energy sources are challenging to integrate, even though they have revolutionized the reduction of environmental consequences and construction-related running costs. Regional and climatic variations, structural and topographical constraints, and political and regulatory barriers commonly impede the use of renewable technologies in buildings. The section that follows goes into greater detail about these limitations and provides information on potential remedies as well as how these challenges could affect projects.

The largest barriers to integrating renewable energy technologies into construction projects are the structural and spatial limitations imposed by available space. Most buildings don't have the surface area or structural strength required for renewable energy installations, especially in crowded urban areas. The added weight of renewable energy systems, such as wind turbines, solar panels, and geothermal loops, must be supported by buildings; many older structures were not built to withstand them. This leads to very costly reconfigurations or even disqualification of the entire refit. For example, the weight of a building's frame is greatly increased by rooftop solar panels. If the roof is not constructed to withstand weight, reinforcement is required, increasing the complexity and cost. Similarly, wind turbines generate dynamic stresses that can strain a building's structural stability, especially in towering buildings. The rooftop and ground space of urban buildings is insufficient for the installation of extensive renewable systems. Wide, interference-free spaces are generally preferred for solar installations.

TABLE 2.7

Leading Energy Modelling Tools

Tool Name	Developer	Supported Platforms	Renewable Simulation Capabilities	Climate Sensitivity	Interoperability (BIM/CAD)	Suitable Project Scale
EnergyPlus	DOE (USA)	Windows/Linux	PV, HVAC, wind, daylighting	High	IFC, OpenStudio	All scales
DesignBuilder	DesignBuilder Ltd.	Windows	PV systems, lighting, occupancy loads	High	AutoCAD, Revit	Medium to large
TRNSYS	TESS	Windows	Solar thermal, PV, geothermal	Medium	Custom modules	Research, commercial
IES VE	IES Ltd.	Windows/macOS	Renewable energy forecasting, LEED	Very high	Revit, gbXML	Medium to large
OpenStudio	NREL	Windows/Linux/macOS	Plugin-based renewable modelling	High	EnergyPlus API	Residential, small
HOMER Pro	HOMER Energy	Windows	Hybrid systems: solar + wind + storage	Medium	Limited	Microgrids, rural

In highly populated metropolitan regions, however, the generation of any solar system is significantly impacted by extra building shadows.

> *Case Study:* Cities like Tokyo or New York that have a limited amount of rooftop space and shadow need to think outside the box. Some creative solutions include mounting solar panels on vertical facades or putting in place community solar projects that produce electricity off-site. It can be challenging to bury loops for geothermal heating and cooling systems in areas with limited acreage or bad soil. Vertical boreholes save space, but because of underground services and costly installation, they might be problematic in metropolitan areas. In certain places, the adoption of renewable energy systems may be limited due to the significant impact that local climatic and geographic circumstances have on their viability.

The ideal settings for solar energy systems are deserts and tropical regions with lots of sunlight. However, in regions with long winters, high cloud covers, or limited daylight, solar panels may not generate enough energy to meet building demands. For example, Scandinavian countries struggle to run their solar systems during the winter months when the days are quite short, despite their tremendous efforts to generate renewable energy. Usually, they supplement it with complementary systems like wind energy or energy storage. A steady wind flow is necessary for wind turbines to operate effectively. In locations with weak or variable wind patterns, turbines might not provide enough energy to be erected. In urban environments, turbulent wind flows caused by buildings and other obstructions significantly reduce turbine efficiency.

> *Case Study:* Wind energy systems are not practicable in Singapore due to the country's very low wind speeds. To achieve its targets for renewable energy consumption, the city-state instead concentrates on solar and waste-to-energy technologies. Subterranean elements like soil type, rock composition, and groundwater availability influence the suitability of geothermal systems. Hard bedrock or low heat conductivity may lead to poorer system performance or increased installation costs. For example, Iceland's volcanic activity has resulted in an abundance of geothermal resources. In most other locations with subpar endowments, the cost of geothermal systems is prohibitive.

Other renewable energy facilities may potentially be destroyed by weather-related calamities like hurricanes, intense snowfall, or heat waves. Wind turbines may need to be turned off during storms, solar panels may be damaged by hail, and geothermal systems may be disrupted by earthquakes or flooding. Government rules and policies play a major role in the adoption of renewable energy technology. While disparate policies and regulatory barriers more often obstruct progress, supportive policies speed up their unification. Building renewable energy systems usually requires complicated clearance and approval processes, which can increase project costs and cause delays. Due to regional variations in zoning laws, construction ordinances, and environmental impact assessments, developers face challenges.

For example, most states have many organizations involved in the permitting process for solar systems, which has resulted in delays that discourage potential customers. Simplifying the permitting process, as California's Solar Permitting Efficiency Act does, can address this. The future is unclear for developers and property owners because of inconsistent regional policies. Renewable energy incentives, such as tax credits, rebates, or feed-in tariffs, are frequently subject to changes in the political environment or simply expire without renewal, which disrupts long-term planning.

> *Case Study:* Following the removal of feed-in subsidies, the number of new solar energy installations in the UK fell precipitously in 2019. This illustrates how stable and predictable rules are essential for the adoption of renewable energy.

Most construction codes are out of date and do not keep up with emerging renewable energy sources. Even if the renewable projects might satisfy modern sustainability standards, this discrepancy usually makes them in violation. For example, most building codes in many nations forbid the incorporation of renewable energy sources, such as solar panels or geothermal systems, into new constructions. California's recent law requiring solar panels on all new residential structures is an example of how code amendments can promote extra growth, and more progressive building laws better highlight this point.

There are several important problems with grid connectivity when it comes to renewable energy systems that generate electricity. The inability of the grid infrastructure to control the distributed energy input usually results in inefficiency or instability. Further, in certain places, insufficient or nonexistent net metering regulations, which allow building owners to sell the excess energy generated back to the utility grid, reduce the financial viability of such renewable systems. For example, India's inconsistent net metering laws and insufficient grid capacity have prevented rooftop solar systems from becoming widely used there.

The challenges of incorporating renewable energy into buildings will require a multipronged approach that combines ingenuity, collaboration, and supportive legislation. Technological developments also reduce spatial and structural constraints. For instance, lightweight solar panels, VAWTs, and small geothermal systems all are in various stages of development and can be applied to a range of building types and environments. Transparent solar panels that can be used in place of windows and solar-integrated construction materials that optimize space efficiency are two

TABLE 2.8

System Cost, Payback Time, and Long-Term Savings

Project Type	Average Installation Cost ($/kW)	Average Payback Period (Years)	20-Year Savings ($/kW)	Maintenance Cost (% of CapEx/Year)	ROI over 20 Years (%)
Small residential	2,700	8.5	9,000	1.5%	230%
Urban mid-rise office	1,800	6.2	18,000	1.2%	390%
Industrial warehouse	1,350	4.8	25,000	1.0%	520%
Educational campus	1,700	6.0	22,000	1.3%	460%
Mixed-use development	2,200	7.2	15,500	1.4%	310%

more innovative ideas. This ensures that the renewable systems are effective by adjusting them to the local environment. Hybrid systems that use wind or geothermal energy balance out energy generation in areas with minimal sunlight. When renewable energy sources aren't providing as much, energy storage steps in to fill the void.

Modern building regulations, expedited approval processes, and ongoing incentives are necessary to remove regulatory barriers. Governments should also invest in grid infrastructure and implement favourable net metering policies to support distributed energy systems. International collaboration and knowledge sharing can help adopt best practices in the integration of renewable energy. Public knowledge and involvement are crucial to overcoming the resistance to renewable energy initiatives. Stakeholder conversations, instruction, and clear planning processes will calm concerns and provide the necessary support. Table 2.8 assesses how system cost-effectiveness changes with project size, comparing small residential, mid-size commercial, and large-scale industrial buildings.

2.7 Future Directions for Renewable Energy in Construction

Global sustainability goals will be aided by creative, technologically sophisticated, and policy-supportive renewable energy in building. Emerging technology, smart energy management systems, and sensible policy frameworks will bring about a significant transformation in building design, construction, and operation. The advancement of renewable technology can lead to increased building flexibility and energy efficiency. Transparent PV panels fitted in windows and facades enable the production of power and natural illumination. Innovations like solar paints, which convert sunlight into energy, and perovskite solar cells, which are much lighter and less expensive, have the potential to radically alter the paradigm of solar integrations. Vertical wind turbines and floating solar farms are renewable energy options in places with limited space or that are very crowded due to urbanization. The use of energy management indoors to better optimize energy production, storage, and consumption is rapidly changing due to smart grids and artificial intelligence. For optimal efficiency, AI systems can examine data on weather trends, building usage, and renewable energy sources. The integration of dispersed energy resources, such as solar panels and wind turbines, is further improved by smart grids' ability to communicate in both directions between buildings and energy providers.

Incorporating renewable energy into building design requires a structured and iterative decision-making process that accounts for site conditions, energy goals, architectural constraints, and economic feasibility. From conceptual planning to final implementation, each phase involves multidisciplinary collaboration and simulation-based analysis. Figure 2.9 presents a comprehensive flowchart outlining the integration process of renewable energy in building design. It details the logical steps from feasibility studies and modelling to component selection and commissioning illustrating how

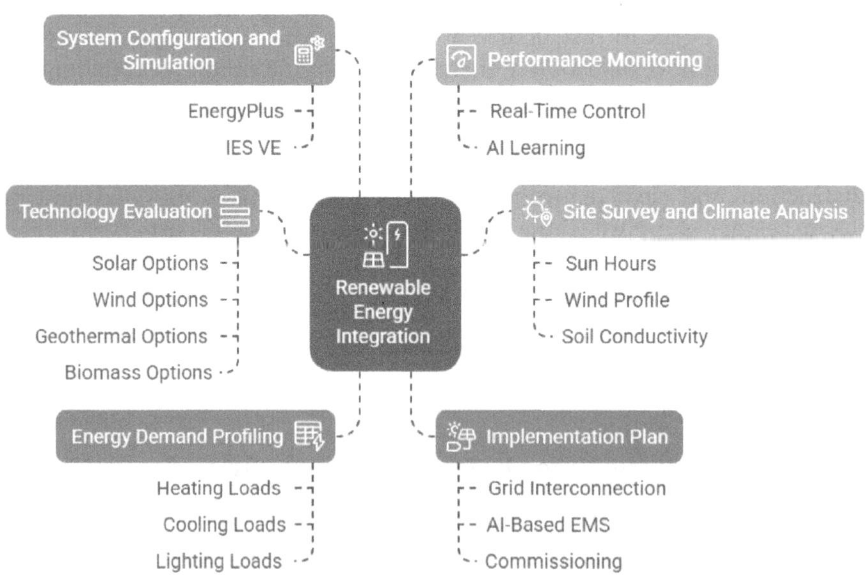

FIGURE 2.9
Flowchart of renewable energy integration in building design.

TABLE 2.9
Emerging R&D Domains in Renewable Construction Technology

R&D Focus Area	Description	Current Stage	Projected Breakthrough Year
Solar-integrated materials	PV glass, solar tiles	Emerging	2028
AI energy management	Predictive load balancing, self-learning grids	Early adoption	2027
3D-printed solar structures	Printed walls with embedded cells	Experimental	2030
Smart geothermal networks	Grid-tied underground systems	Pilot testing	2029

sustainable energy solutions are woven into the architectural and engineering workflow from the outset.

AI-powered microgrids, for example, might improve energy resilience by storing excess energy from intermittent renewable sources and distributing it when demand is higher. For the use of renewable energy in construction to increase more quickly, supportive legislation will be required. The government can expedite license acquisition, provide direct financial incentives, and require renewable systems in new development, as discussed in Table 2.9. By providing subsidies for innovative technologies, policies that facilitate research and development will also be promoted. In order to secure the construction industry's future, international cooperation regarding green building certification and standards will further stimulate complete implementation of renewable technology.

2.8 Conclusion

The primary route to sustainable development is the use of renewable energy sources in buildings. Energy modelling software, appropriate material choices, and architectural concepts for integrating renewable systems are some of the major design aspects covered in this chapter. Also, it has demonstrated several feasible uses for geothermal, wind, and solar energy in infrastructure, commercial, and residential projects. A smaller carbon footprint, urban sustainability, and compelling proof that renewable energy sources reduce greenhouse gas emissions to slow down climate change are a few of these advantages. Energy resilience, increased property values, and life cycle cost savings are economic aspects that support the value of renewable technologies in construction. Among the challenges identified were structural constraints, weather variations, and regulatory barriers; these necessitated innovative solutions and supportive legislation. The role of AI and smart grids, as well as emerging technologies and policy recommendations, were then leveraged to provide more direction on how to accelerate adoption and increase system efficiency. The building industry is well positioned to lead this transformation. Renewable technology-based designs reduce their negative environmental consequences, increase energy efficiency, and help achieve global sustainability goals. This suggests that collaboration between architects, structural engineers, developers, and policymakers is vital to removing barriers to the rapid scaling of such renewable energy solutions. A clear strategy with consistent incentives and permitting processes would spur more innovation and broader use. It is more crucial than ever to look for cutting-edge technological developments in the areas of urbanization and resource scarcity, such as AI-powered energy management systems and emerging renewable technologies. A sustainable built environment, reduced dependency on fossil fuels, and urban environment transformation all are possible with the use of renewable energy in construction. The focus on incorporating renewables will make future construction projects energy-efficient, environmentally responsible, and economically advantageous over the long run. In this way, it paves the way for a strong and sustainable future for the construction industry and the communities it serves.

Chapter Review: Chapter 2 – Renewable Energy and Sustainable Building Systems

Multiple Choice Questions

Test your understanding of the key concepts from this chapter. Each question is followed by four options, only one is correct.

1. What is the main goal of smart building energy management systems?
   A. Reducing the cost of construction
   B. Automating design decisions
   C. Optimizing energy use and occupant comfort
   D. Increasing property size
2. Circular economy in construction promotes:
   A. Linear resource consumption
   B. Demolition over reuse
   C. Recycling and closed-loop material flows
   D. Increased fossil fuel usage

3. The 30-year evolution of energy management in buildings highlights:
   A. A decline in building complexity
   B. Increasing reliance on manual systems
   C. A shift to data-driven and automated solutions
   D. Reduction in sustainability concerns
4. The smart energy management funnel is used to:
   A. Increase heating costs
   B. Eliminate user inputs
   C. Narrow energy use insights for efficient control
   D. Expand fossil fuel use
5. A geothermal heat pump system works by:
   A. Drawing heat from solar panels
   B. Using groundwater to power turbines
   C. Transferring heat to and from the earth
   D. Burning biofuels underground
6. One major advantage of renewable energy in construction is:
   A. Higher carbon emissions
   B. Constant reliance on the power grid
   C. Reduced operational costs and emissions
   D. Decreased design flexibility
7. Sustainable building practices have evolved to include:
   A. Focus only on aesthetics
   B. Passive design without analytics
   C. Smart sensors and data-driven models
   D. Manual lighting and ventilation
8. Integrated renewable energy systems help buildings by:
   A. Increasing grid dependency
   B. Creating isolated power zones
   C. Balancing energy production, storage, and use
   D. Replacing all electronics
9. Flowcharts for renewable integration aid in:
   A. Promoting fossil-based systems
   B. Unstructured energy planning
   C. Systematic design and decision-making
   D. Eliminating design constraints

10. Circular economy reduces environmental impact by:
    A. Encouraging new raw material extraction
    B. Sending waste to landfills
    C. Reusing construction materials
    D. Increasing demolition frequency
11. A key benefit of digital twins in energy management is:
    A. Ignoring sensor feedback
    B. Simulating and optimizing performance in real time
    C. Replacing design teams
    D. Removing energy systems
12. Building energy evolution is characterized by:
    A. Static technologies
    B. Fully manual control
    C. Growing intelligence in systems
    D. Declining automation
13. Smart energy systems use sensors to:
    A. Increase randomness in data
    B. Disable automation
    C. Monitor environmental variables for control
    D. Remove thermal insulation
14. Human roles in Construction 5.0 are shifting towards:
    A. Manual calculations
    B. Supervisory and strategic decision-making
    C. Ignoring AI-based recommendations
    D. Repetitive energy logging
15. Renewable integration in buildings requires:
    A. A focus only on aesthetic features
    B. Ignoring site-specific factors
    C. Feasibility analysis and life cycle planning
    D. One-size-fits-all energy systems

Answer Key

Question No.	Correct Option
1	C
2	C
3	C
4	C
5	C
6	C
7	C
8	C

Question No.	Correct Option
9	C
10	C
11	B
12	C
13	C
14	B
15	C

References

Ali, I. H. O., Ouassaid, M., & Maaroufi, M. (2021). Optimal appliance management system with renewable energy integration for smart homes. In *Renewable energy systems* (pp. 533–552). Elsevier.

Ansari, M. F., Kharb, R. K., Luthra, S., Shimmi, S., & Chatterji, S. (2013). Analysis of barriers to implement solar power installations in India using interpretive structural modeling technique. *Renewable and Sustainable Energy Reviews, 27,* 163–174.

Aresti, L., Alvi, M. R., Cecinato, F., Fan, T., Halaj, E., Li, Z., Okhay, O., Poulsen, S. E., Quiroga, S., & Suarez, C. (2024). Energy geo-structures: A review of their integration with other sources and its limitations. *Renewable Energy, 230,* 120835.

Baral, A., & Guha, G. S. (2004). Trees for carbon sequestration or fossil fuel substitution: The issue of cost vs. carbon benefit. *Biomass and Bioenergy, 27*(1), 41–55.

Biyik, E., Araz, M., Hepbasli, A., Shahrestani, M., Yao, R., Shao, L., Essah, E., Oliveira, A. C., Del Cano, T., & Rico, E. (2017). A key review of building integrated photovoltaic (BIPV) systems. *Engineering Science and Technology, an International Journal, 20*(3), 833–858.

Butt, T., Braima, M., Maraqa, M., Goodhew, S., Sundaram, S., & Mohamed, M. (2024). A perspective on solar railway potential for the UK. *Environmental Progress and Sustainable Energy, 43*(6), 1–10.

Debbarma, M., Sudhakar, K., & Baredar, P. (2017). Comparison of BIPV and BIPVT: A review. *Resource-Efficient Technologies, 3*(3), 263–271.

Dincer, F. (2011). The analysis on photovoltaic electricity generation status, potential and policies of the leading countries in solar energy. *Renewable and Sustainable Energy Reviews, 15*(1), 713–720.

Eichholtz, P., Kok, N., & Quigley, J. M. (2010). Doing well by doing good? Green office buildings. *American Economic Review, 100*(5), 2492–2509.

Galantino, C. R., Beyers, S., Anderson, C. L., & Tester, J. W. (2021). Optimizing Cornell's future geothermal district heating performance through systems engineering and simulation. *Energy and Buildings, 230,* 110529.

Gerber, L., & Maréchal, F. (2012). Defining optimal configurations of geothermal systems using process design and process integration techniques. *Applied Thermal Engineering*, *43*, 29–41.

Guamán-Rivera, R., Martínez-Rocamora, A., García-Alvarado, R., Muñoz-Sanguinetti, C., González-Böhme, L. F., & Auat-Cheein, F. (2022). Recent developments and challenges of 3D-printed construction: A review of research fronts. *Buildings*, *12*(2), 229.

Huang, L., Krigsvoll, G., Johansen, F., Liu, Y., & Zhang, X. (2018). Carbon emission of global construction sector. *Renewable and Sustainable Energy Reviews*, *81*, 1906–1916.

Ibn-Mohammed, T., Greenough, R., Taylor, S., Ozawa-Meida, L., & Acquaye, A. (2013). Operational vs. embodied emissions in buildings: A review of current trends. *Energy and Buildings*, *66*, 232–245.

Kumar, R., Raahemifar, K., & Fung, A. S. (2018). A critical review of vertical axis wind turbines for urban applications. *Renewable and Sustainable Energy Reviews*, *89*, 281–291.

Lai, C. S., Jia, Y., Lai, L. L., Xu, Z., McCulloch, M. D., & Wong, K. P. (2017). A comprehensive review on large-scale photovoltaic system with applications of electrical energy storage. *Renewable and Sustainable Energy Reviews*, *78*, 439–451.

Maglad, A. M., Houda, M., Alrowais, R., Khan, A. M., Jameel, M., Rehman, S. K. U., Khan, H., Javed, M. F., & Rehman, M. F. (2023). BIM-based energy analysis and optimization using insight 360 (case study). *Case Studies in Construction Materials*, *18*, e01755.

Meinshausen, M., Lewis, J., McGlade, C., Gütschow, J., Nicholls, Z., Burdon, R., Cozzi, L., & Hackmann, B. (2022). Realization of Paris agreement pledges may limit warming just below 2°C. *Nature*, *604*(7905), 304–309.

Moreno Sierra, A. A. (2024). *Technology roadmap of lightweight materials in the energy sector (wind and solar): A comprehensive analysis and strategy for future development*. Universitat Politècnica de Catalunya.

Nørgaard, P., & Poder, S. H. (2019). Perspectives on solar power in dense urban areas: With Copenhagen International School as case study. *The Journal of Engineering*, *2019*(18), 5134–5137.

Sarbu, I., & Sebarchievici, C. (2014). General review of ground-source heat pump systems for heating and cooling of buildings. *Energy and Buildings*, *70*, 441–454.

Stober, I., & Bucher, K. (2013). *Geothermal energy* (pp. 973–978). Germany: Springer.

Tang, C. S., & Zhou, S. (2012). Research advances in environmentally and socially sustainable operations. *European Journal of Operational Research*, *223*(3), 585–594.

Weckend, S., Wade, A., & Heath, G. A. (2016). *End of life management: Solar photovoltaic panels*. International Renewable Energy Agency.

Wilkinson, S. J., & Sayce, S. (2020). Decarbonising real estate: The evolving relationship between energy efficiency and housing in Europe. *Journal of European Real Estate Research*, *13*(3), 387–408.

Xu, W., Li, Y., Li, G., Li, S., Zhang, C., & Wang, F. (2021a). High-resolution numerical simulation of the performance of vertical axis wind turbines in urban area: Part II, array of vertical axis wind turbines between buildings. *Renewable Energy*, *176*, 25–39.

Xu, W., Li, G., Zheng, X., Li, Y., Li, S., Zhang, C., & Wang, F. (2021b). High-resolution numerical simulation of the performance of vertical axis wind turbines in urban area: Part I, wind turbines on the side of single building. *Renewable Energy*, *177*, 461–474.

3

Renewable Energy and Industry 4.0

3.1 Overview of Integrating Renewable Energy Systems with Industry 4.0

Technology has caused a substantial shift in how industries think about sustainability, efficiency, and resource management. The Fourth Industrial Revolution, which encompasses well-known digital technologies like Internet of Things (IoT), artificial intelligence (AI), blockchain, big data analytics, and robotics, is now referred to as Industry 4.0. Since integrated systems allow for real-time data processing for informed decision-making, all of this is creating changes in a lot of firms. Since the construction industry is one of the biggest contributors to carbon dioxide emissions and global energy consumption, the pertinent technologies in this field have the potential to significantly disrupt energy production, distribution, and consumption, especially in the renewable energy sector.

3.1.1 Industry 4.0 Technologies Overview

The automation and digitization of traditionally carried out industrial operations are the main emphasis of Industry 4.0, which revolves there. Connecting systems, sensors, and gadgets that can gather and exchange data is made possible by the IoT concept. Such a network can offer real-time monitoring and control of the best energy system operating. As a result, IoT-based solar panels and wind turbines would be able to track energy output and the environment in the hopes of increasing efficiency and lowering maintenance costs. According to IDC, the energy sector's use of IoT will only increase, reaching $31 billion by 2023 (Albreem et al., 2023). The energy sector can leverage the power of machine learning, natural language processing, and predictive analytics to enable efficient forecasting, supply optimization, and the identification of anomalies in energy system failure. AI is the other crucial component of Industry 4.0. Applications backed by AI algorithms, such as renewable energy forecasting, improve forecast accuracy by 20–30% (Ahmed & Khalid, 2019). Integrating variable sources like solar and wind power into the system is necessary to maintain stability and dependability. Blockchain technology

DOI: 10.1201/9781003599272-3

enables all of this, resulting in a secure, decentralized energy trading system. Direct sales of excess energy from consumers to other users are made possible by P2P energy trading platforms' blockchain technology. In a case study based on the Australia Power Ledger programme, researchers demonstrated that blockchain-enabled energy trading reduced electricity prices by 25% (Ali et al., 2023). *Digital Twin:* Virtual duplicates of physical systems have made it possible to simulate and optimize energy systems more effectively than ever before, even before they are implemented. Engineers may predict potential inefficiencies, simulate the result, and implement corrective measures using these computer models. A McKinsey & Company analysis claims that by lowering operational expenses, the use of a digital twin in energy management boosts system efficiency by 20–30% (Borowski, 2021).

3.1.2 Industry 4.0 and Renewable Energy Convergence

The integration of Industry 4.0 technology with renewable energy sources will be crucial to advancing the energy transformation process. Conventional energy production methods are costly, unclean, and ineffective. Thus, the transition will be accelerated against challenges for renewable energy systems with the rapid advancement of digital tools in cleaner and sustainable sources. The ability to track energy generation and use in detail on the IoT is a key component of this transition. Therefore, by integrating IoT devices and balancing supply and demand in real time, smart grids ensure the optimum energy distribution techniques. Further, according to the European Commission's Joint Research Centre, smart grids enabled by the IoT have helped to reduce power outages by 20% and increase energy efficiency by 15% (Kabalci et al., 2019). The ability of AI to examine data from these IoT devices for patterns that could be utilized to predict future energy requirements and enhance system efficiency is another aspect of this synergy's power. For example, AI-based energy management in solar farms increased energy production by up to 10%. Also, it offers predictive maintenance that decreases downtime and increases the lifespan of renewable energy infrastructure (Onwusinkwue et al., 2024). Blockchain makes it possible for local communities to generate, store, and exchange renewable energy in a decentralized energy system, democratizing energy access and granting them more autonomy from conventional energy suppliers. A fine example of this kind of organization is the Brooklyn Microgrid, a blockchain-based initiative in the United States where locals trade extra solar energy and report a 40% decrease in carbon emissions (Boumaiza & Maher, 2024). To appreciate the scale of transformation driven by Industry 4.0 in renewable energy projects, it's crucial to examine global adoption trends. As shown in Table 3.1, technologies such as IoT and digital twins have seen rapid uptake in developed economies due to supportive infrastructure and policy frameworks. AI adoption is also increasing steadily, especially in Asia-Pacific, where data-driven construction optimization is being widely explored.

TABLE 3.1

Comparative Analysis of Traditional versus Smart Energy Systems in Construction

Metric	Traditional System	Smart System (Industry 4.0)	% Improvement	Cost Impact	Energy Loss (%)	Downtime (h/year)	Monitoring Capability
Energy efficiency (%)	60	85	+41.7%	↓ OPEX	15	200	Manual
Data accessibility	Low	High	+300%	↑ CAPEX	N/A	N/A	Real-time
Maintenance cost ($/year)	$120K	$85K	–29.2%	↓ OPEX	N/A	N/A	Predictive
Response time (h)	10	2	–80%	↓ Downtime	N/A	N/A	Automated
Decision support	None	AI driven	Infinite	↑ Productivity	N/A	N/A	Smart
Carbon footprint (tCO_2/year)	500	300	–40%	↓ Emissions	N/A	N/A	Auditable
Integration with renewables	Low	High	+250%	↑ ROI	N/A	N/A	Seamless

3.1.3 Applicability to Construction Projects

The construction industry, which uses 36% of the world's energy and produces over 40% of its CO_2 emissions, is perhaps the one that stands to gain the most from a combination of Industry 4.0 technologies and renewable energy. Technology-enabled smart construction techniques can offer a way to improve project results, lessen environmental impact, and increase energy efficiency. Real-time energy system monitoring and control are already made possible by the incorporation of IoT devices into the building construction process. A building equipped with sensors, for example, will be able to pinpoint its energy use, areas for efficiency improvement, and recommendations that are likely to boost performance. In a National Renewable Energy Laboratory trial, for instance, an IoT-based energy management system cut energy use in commercial buildings by 25% (Saleem et al., 2023). AI-powered smart construction technologies streamline processes and enhance decision-making. Algorithms based on machine learning that analyse past trends in energy use forecast future demand and assist in the construction of new, energy-efficient buildings. AI-driven design software assists architects in creating layouts that use less energy. According to an American Institute of Architects study, buildings that incorporate AI into building design frequently save 20% on energy costs (Mehmood et al., 2019). It enables the development of

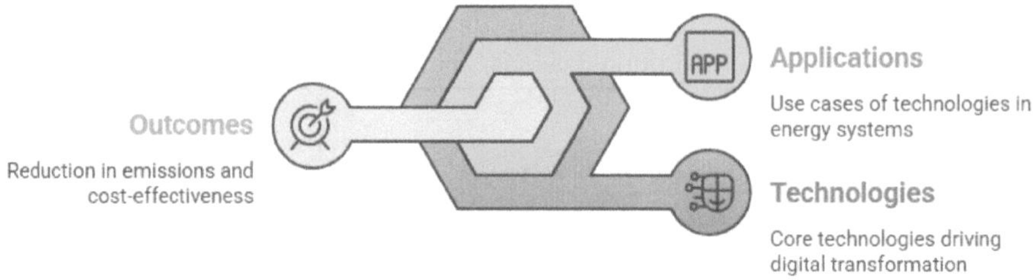

FIGURE 3.1
Digital transformation in renewable energy.

digital twins and aids in the planning and execution of building projects. Virtual models of buildings and energy systems can be used to evaluate different scenarios and difficulties, as well as implement solutions, before the actual process begins. This method guarantees sustainable criteria, lowers the cost, and shortens the time frame. More than 75% of major construction businesses will have used digital twins to enhance project outcomes by 2027 (Afzal et al., 2023). Blockchain technology has the potential to transparently facilitate decentralized energy management and supply chain processes in construction projects. For example, blockchain technology can be used to further verify the sustainability of building materials by tracking their origin. Technology makes sure that vital services like hospitals and emergency services continue to receive supplies even in the event of supply disruption, which further lowers the danger of any grid-related fire threats. These initiatives strengthened the system's resilience and demonstrated how the smart grid may be modified to address problems that the energy paradigm is unable to address. However, the advantages of flexible smart networks go beyond their technological effectiveness. For example, decarbonizing energy use makes it easier to integrate renewable energy sources into the fight against climate change. To encapsulate the synergy among digital tools, Figure 3.1 presents a node-based network map showing interconnectivity between systems. For example, AI connects to IoT for sensor data processing, to blockchain for secure transaction handling, and to digital twins for simulation feedback.

3.2 Smart Technologies in Energy Management

This is the key to the revolution in energy management since it makes it simpler to monitor, control, and optimize energy resources, especially those connected to renewable energy systems. The new technologies improve sustainability and efficiency. Examples of this type of technology include digital twins, AI, and the IoT. Each has a distinct function in energy-related operations and offers a special contribution to a more effective strategy for promoting sustainability at higher levels.

THOUGHT BOX

DIGITAL TWIN VERSUS BIM: WHAT'S THE DIFFERENCE?

While building information modelling (BIM) focuses primarily on static models used during the design and construction phases, a *digital twin* represents a *living, continuously updated* replica of a building system throughout its life cycle. Unlike BIM, digital twins incorporate *real-time data* through sensors and IoT to *simulate, predict,* and *optimize* operations. Think of BIM as the blueprint, and the digital twin as the brain that keeps learning. Which would you prioritize in a smart energy project?

3.2.1 IoT in Renewable Energy: Energy System Control and Real-Time Monitoring

Interconnected networks of sensors and equipment are being prioritized by the IoT, which is reengineering the management of renewable energy sources. A renewable energy system would be more reliable and efficient if IoT-enabled equipment produced real-time data on energy generation, distribution, and consumption. Solar panels, wind turbines, energy storage devices, and many other renewable energy–related items can be monitored by the IoT. IoT sensors gather information on weather, energy production, and device operation so that operators can take prompt corrective action. IoT sensors, for instance, track the operation of individual solar panels to pinpoint malfunctioning parts and reduce energy loss. According to a 2022 report by research firm Grand View Research, there is a significant demand for the global IoT in energy, which is predicted to grow at a 17.1% CAGR from 2023 to 2030 (Singh & Sehgal, 2024). In addition to monitoring, IoT makes it possible to manage installations of renewable energy. For example, smart grids employ IoT to dynamically balance energy supply and demand to maximize the usage of renewable resources. Turbines may be positioned in real time towards the most advantageous wind direction and speed, thanks to IoT applications in the wind energy sector. IoT also facilitates predictive maintenance. Predictive maintenance saves money on maintenance and downtime by preventing equipment failure before it really happens. Predictive maintenance of renewable energy plants is expected to reduce operating costs by about 25% (Turnbull & Carroll, 2021). Efficiency in energy distribution and storage is another guarantee that comes with integrating renewable energy sources. Beyond basic metering, today's intelligent buildings leverage IoT-enabled systems to form closed feedback loops that optimize energy behaviour in real time. Figure 3.2 captures the technical underpinnings of

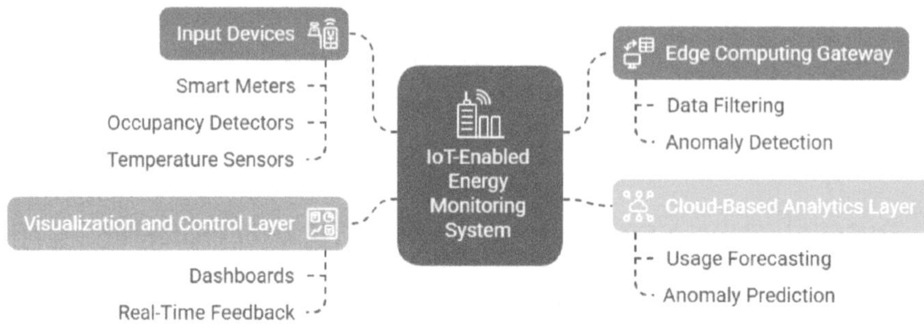

FIGURE 3.2
IoT-enabled energy monitoring system architecture.

this architecture. Starting from the input layer, it details how environmental sensors (measuring temperature, solar irradiance, and humidity) channel data through wireless protocols like ZigBee or LoRaWAN. The transmission layer leads to edge computing modules that preprocess critical data before it's sent to cloud-based analytics engines. Importantly, the system closes the loop by triggering adjustments in heating, ventilation, and air conditioning (HVAC) or lighting systems, with performance metrics tied to latency reduction, power savings, and system responsiveness.

For instance, a case study on IoT-based microgrids showed that they increase the reliability of electricity supplies in rural areas and reduce energy losses by 15% (Silva et al., 2023). Smart construction projects radically differ from traditional counterparts in terms of energy system design and efficiency. Table 3.2 outlines these differences across multiple dimensions, including real-time monitoring, predictive maintenance, energy efficiency, and data transparency, demonstrating how Industry 4.0 technologies enhance operational outcomes in modern green buildings.

3.2.2 AI for Optimization: Energy Efficiency Predictive Analytics

Predictive analytics and machine learning techniques are used by AI to optimize energy systems. These forecasts show how much energy will be needed in the future, spot trends in the massive amounts of data coming from renewable energy sources, and offer recommendations for enhancing their effectiveness. The most important application of AI in relation to renewable energy is energy forecasting. AI-based models provide incredibly precise forecasts for energy output and demand when historical meteorological data, energy consumption patterns, and environmental factors are taken into consideration. For intermittent renewable energy sources like solar and wind, whose energy output fluctuates depending on the weather, this is quite beneficial. AI-based energy forecasting, for instance, decreased forecasting errors by 30%, making it easier to integrate renewable energy sources into the

TABLE 3.2

IoT Device Categories and Their Roles in Renewable Energy Monitoring

Device Type	Role	Data Captured	Connectivity	Example Sensor	Cost Range ($)	Accuracy (%)	Deployment Area
Smart meters	Usage tracking	kWh	Wi-Fi	Zigbee Meter	50–300	98	Residential
Environmental sensors	Condition tracking	Temperature humidity	LoRa	Bosch BME688	100–500	95	Rooftops
Power quality monitors	Grid health	Voltage, frequency	Ethernet	PQube 3	400–2,000	99	Grid nodes
Solar inverters	Conversion/efficiency	PV output	Modbus	Sunny Boy	1,000–3,000	97	Solar Panels
Vibration sensors	Maintenance prediction	Motion	5G	Smart vibe	80–250	92	Turbines
Flow meters	Fluid tracking	L/min	Wi-Fi	ABB digital	500–1200	96	Cooling systems
Digital relays	Switching and protection	Load status	IoT	SEL-751	200–1500	99	Substations
Occupancy sensors	Space management	Motion	Zigbee	Philips Hue	30–100	94	Buildings

grid (Ukoba et al., 2024). As building assets become more complex, AI-driven predictive maintenance has emerged as a game-changer for system longevity and cost control. Figure 3.3 visualizes this intelligent pipeline, starting with anomaly data generated by embedded sensors. This data is then fed into machine learning models such as LSTM neural networks or decision trees which assess the likelihood of system failure. Based on probability thresholds, the system schedules maintenance interventions in advance.

THOUGHT BOX

SHOULD BUILDINGS BECOME MICRO-ENERGY TRADERS?

With the rise of blockchain platforms and smart meters, buildings are no longer just energy consumers; they can also *generate and trade surplus energy*. This brings up crucial questions: Should every building have its own *wallet* and smart contract? How will utility companies adapt when the consumer also becomes the *supplier*? Peer-to-peer (P2P) models challenge traditional centralized control *is the grid ready for this shift*?

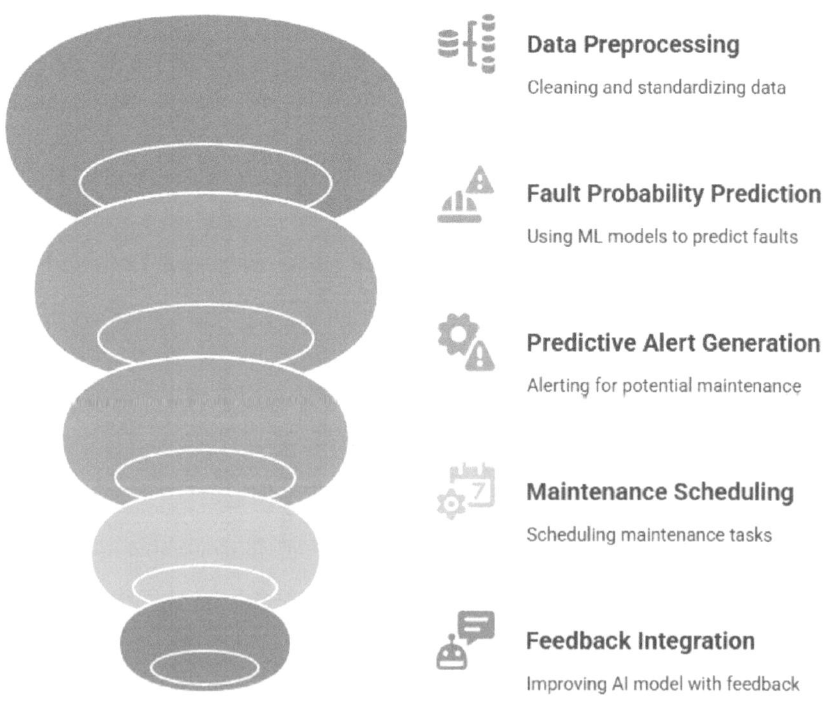

FIGURE 3.3
AI-driven predictive maintenance process.

This is the application of AI-powered predictive maintenance to increase the lifespan and dependability of renewable energy systems. Machine learning algorithms will use data from IoT sensors to identify irregularities, forecast equipment failures, and even suggest remedies. In this manner, early indicators of gearbox wear or stress on wind turbine blades will alert AI before costly malfunctions occur. Research from Deloitte suggests that AI Predictive Maintenance could boost wind turbine efficiency by as much as 10–15% (Chatterjee & Dethlefs, 2021). Apart from optimization and maintenance, AI makes grid management and energy trading easier. The best pricing for trading renewable energy is determined by AI algorithms based on market patterns and energy availability. Smart grids are powered by AI, which permits automatic demand-based modifications and energy flow regulation. For example, the UK test revealed that an AI-enabled management system's energy loss was cut by 18%, increasing energy stability (Rojek et al., 2025). AI plays a pivotal role in predicting energy demands and preventing overproduction or loss. As illustrated in Table 3.3, multiple case studies reveal significant waste reductions ranging from 12% to 38% after the deployment of AI-powered forecasting tools in solar and wind installations (Ukoba et al., 2024). These outcomes underscore the predictive accuracy and financial viability of AI integration.

3.2.3 Digital Twins: Modelling Energy Systems to Gain Operational and Design Understanding

The idea of 'digital twin ideas' has been explored, in which modern energy resources are managed through a virtual representation of real energy systems, with models used to mimic design, operation, and optimization to comprehend how different energy system settings work. Prior to being used in the construction of a renewable project, this study employed a digital twin to examine the energy system. Engineers' use of digital twins to model wind direction, energy output, and even turbine alignment at wind farms is a typical example. It allows designers to make changes to designs with the least amount of energy waste and maximum effectiveness. When comparing legacy construction with smart building approaches, the advantages of digital twins become evident. Figure 3.4 contrasts both life cycles side by side from design and material selection to real-time operation and end-of-life recycling. Digital twins, embedded with IoT feedback, enable simulation-driven optimization at every stage. For instance, predictive modelling can suggest design changes that reduce embodied energy, while dashboards help track carbon outputs in real time.

TABLE 3.3

AI Applications in Energy Optimization: Use Case Matrix

AI Model	Energy Function	Input Data	Algorithm Type	Output	Accuracy (%)	Deployment Cost ($)	Scalability
LSTM RNN	Demand forecasting	Historical load	Deep learning	kWh prediction	96	10,000	High
CNN	Thermal imaging	Infrared data	Supervised	Heat map	93	8,500	Medium
SVM	Fault detection	Vibration, load	Classification	Fault class	90	7,000	Medium
Random forest	Peak load prediction	Real-time metering	Ensemble	Load curve	92	9,000	High
K-means	Asset clustering	IoT device data	Unsupervised	Group ID	88	6,000	High
Bayesian networks	Risk assessment	Maintenance logs	Probabilistic	Risk score	85	4,000	Medium
Deep Q-learning	Grid balancing	Grid feedback	Reinforcement	Action policy	94	12,000	Low
Decision trees	Energy waste tracing	Sensor readings	Supervised	Anomaly alerts	91	5,500	Medium

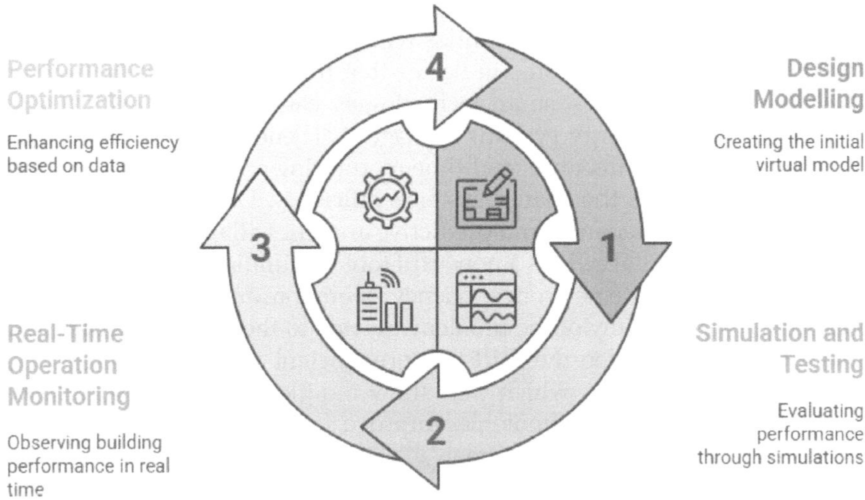

FIGURE 3.4
Digital twin life cycle in sustainable building development.

THOUGHT BOX

THE HUMAN ELEMENT: ARE WE TRAINING ENOUGH DIGITAL-READY ENGINEERS?

Industry 4.0 is as much about *people* as it is about technology. Yet, studies suggest a shortage of professionals trained in both construction and digital systems. Can today's civil engineers interpret AI-driven forecasts? Are project managers trained to analyse data dashboards from IoT? Building future-ready infrastructure requires *digital fluency* across all project roles not just IT departments.

Digital twins can also help with predictive maintenance by modelling the deterioration of energy-producing or energy-transmitting devices. Operators can increase longevity and lost time by using models to properly schedule maintenance procedures. The digital twin of offshore wind farms demonstrates how seawater corrodes certain turbine components, which can be changed as necessary. According to a Norwegian case study, one would anticipate a 15% decrease in turbine downtime and a 10% increase in energy if the digital twin of the turbines is appropriately utilized for efficient maintenance (Alzammar, 2023). As a result, the digital twin principles scale up from a single unit to the entire energy network. These models replicate intricate smart grid and microgrid systems for energy-intensive performance enhancement in addition to unit-level optimization. The integration of renewable energy sources, the effects of energy storage devices, and grid optimization tactics all can be modelled by a digital twin of a city's energy infrastructure. According to a report by Gartner, 75% of utilities will employ energy management by 2027 (Shehabi et al., 2024). This indicates that their significance is growing. The IoT, one of the newest smart technologies, can have the most influence on how renewable energy systems manage their energy, along with digital twins and AI. In this instance, real-time monitoring and control powered by IoT further improves the energy system's efficiency. Demand-side management, efficient maintenance, and predictive analytics all would be made possible by AI. The digital twin is a powerful tool for simulating energy systems and offers insights into network efficiency, operation, and design. Since these technologies solve many of the problems related to the integration of renewable sources, they hasten the shift to more efficient and sustainable energy management techniques, which will allow the global sustainability movement to thrive. As these technologies improve, they will undoubtedly have a considerably greater impact on renewable energy and energy management, creating a cleaner and more efficient energy future.

3.3 Enhancing Energy Efficiency in Buildings

Building energy efficiency is crucial to achieving global sustainability goals, according to the International Energy Agency, since buildings account for almost 40% of worldwide energy consumption and 28% of energy-related CO_2 emissions (Nejat et al., 2015). The emergence of peer-to-peer renewable energy markets signifies a shift in how energy is consumed and monetized in the built environment. Figure 3.5 illustrates the blockchain-backed framework that supports this paradigm. Each node represents either a producer, consumer, or prosumer, connected via smart meters to a decentralized ledger. Transactions governed by smart contracts are validated through consensus mechanisms. The flow diagram includes token-based credit systems and demand-driven pricing models, clarifying how trust and transparency are built into autonomous energy exchange systems.

3.3.1 Automated Energy Distribution Using Industry 4.0

The majority of modern devices are driven by various machine learning algorithms that depend on analytics to operate. In automated systems, the algorithms may therefore regulate and guide energy distribution without resulting in energy losses. Moving electricity from non-essential parts of a structure to high-demand portions is one of the many activities made possible by automation and AI. We can refer to these technologies as Industry 4.0. Due to the facility's varying average energy consumption, these occupations are regularly changed. The life cycle benefits of digital twin technology

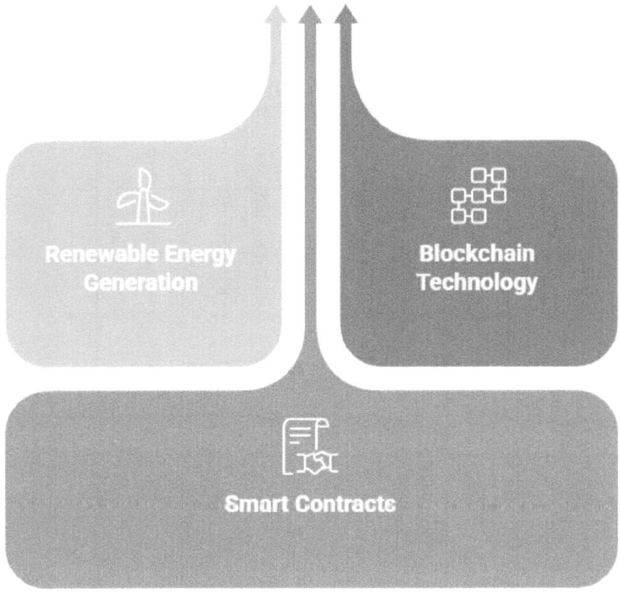

FIGURE 3.5
Decentralized energy trading.

are now well-documented. Table 3.4 presents a breakdown of its applications from design simulation to maintenance scheduling and quantifies advantages such as reduced downtime and optimized HVAC efficiency. The data supports the growing case for digital twin deployment in large-scale construction.

3.3.2 Real-Time Smart Energy Distribution

Real-time data from sensors and IoT-enabled devices is the foundation of energy distribution automation. These gadgets continuously track the trend in energy consumption to assist these systems in efficiently distributing energy. For example, smart HVAC systems employ weather forecasting and real-time occupancy data to function. According to a 2022 study at the Lawrence Berkeley National Laboratory, intelligent autonomous HVAC systems may be able to reduce energy consumption by 30% without compromising occupant comfort (Kong et al., 2022). Automating a building's lighting system is another way that automation can be beneficial. Electricity is saved. Motion sensors and daylight integration technologies are used by automatic controls to turn lights on only when necessary and off when there is an abundance of natural light. The US Department of Energy estimates that automated lighting control systems can save up to 40% for commercial buildings (Lowry, 2016).

TABLE 3.4

Blockchain Use Cases in Distributed Energy Trading

Use Case	Smart Contract Type	Participants	Energy Volume (kWh)	Transaction Speed (s)	Fees (%)	Platform	Regulatory Status
Peer-to-Peer microgrid	Escrow	Households	10–500	10–15	0.5	PowerLedger	Trial
Tokenized solar credit	ERC-20	Producers	50–1,000	8–10	1.2	WePower	Active
Battery swapping	Lease-based	EV owners	5–50	12	0.8	Energy Web	Pilot
Renewable NFT certificates	Identity	Certifiers	N/A	5	2.5	SunExchange	Active
Carbon offset exchange	Compliance	Firms	CO_2 Units	20	1	KlimaDAO	Limited
Grid ancillary services	Oracle-Based	Utilities	100–2,000	15	1	EWF	Pilot
Energy aggregation pools	Delegated	Producers	200–10,000	18	0.7	GridSingularity	Developing
Load forecast bidding	Auction	Prosumer-Gov	Forecast data	22	1.1	VoltMarkets	Trial

3.3.3 Integration with Systems of Renewable Energy

One of the most creative applications of automated energy distribution is in conjunction with renewable energy sources, which is equally astounding. Buildings with solar panels, wind turbines, or energy storage can effectively manage their resources and transition to renewable energy sources during periods of high output before switching back to stored energy or grid power, thanks to automated distribution systems. It lowers energy expenses and reliance on fossil fuels. A net-zero energy building with an automated energy distribution system was the subject of a case study in Germany. In comparison to a traditional system, these energy distribution technologies, integrated solar panels, and battery storage achieved 45% reduced reliance on the grid and 25% higher energy efficiency. These illustrations demonstrate how Industry 4.0 technologies can revolutionize building energy management (Mohamed et al., 2019).

3.3.4 Integration of Sensors and Smart Meters

Installing sensors and smart meters in an Industry 4.0–enabled building results in increased energy efficiency. Since they supply the data needed for real-time monitoring and enhancements, these have long been referred to as the 'eyes and ears' of any cutting-edge energy management system. Real-time recording of vital characteristics, such as occupancy, temperature, humidity, and energy consumption, will be made possible by sensors. Thus, this will serve as the foundation for BMS's well-informed recommendations regarding the locations of important energy-saving controls. To maintain balance in an increasingly decentralized grid, smart grids must be equipped with multilayered intelligence. Figure 3.6 outlines this complexity, starting from renewable generation points like wind and solar farms, flowing through AI-based control centres that predict load requirements, and down to smart substations with autonomous switching mechanisms. At the end-user level, demand response meters adjust usage based on grid conditions. Each system tier is annotated with data flow paths, latency expectations, and reliability metrics depicting the full orchestration needed for a responsive and resilient energy grid.

Making a distinction when heat generation is active and the ambient temperature has risen to its threshold is one example. These features can also benefit larger commercial buildings, whose energy needs may vary significantly throughout the day. The deployment of blockchain in the energy sector is catalysing peer-to-peer trading models, enabling decentralized markets. Table 3.5 lists prominent global pilot projects and start-ups leveraging blockchain for local energy exchanges. Metrics such as transaction latency, cost savings, and user scalability highlight the disruptive potential of these systems.

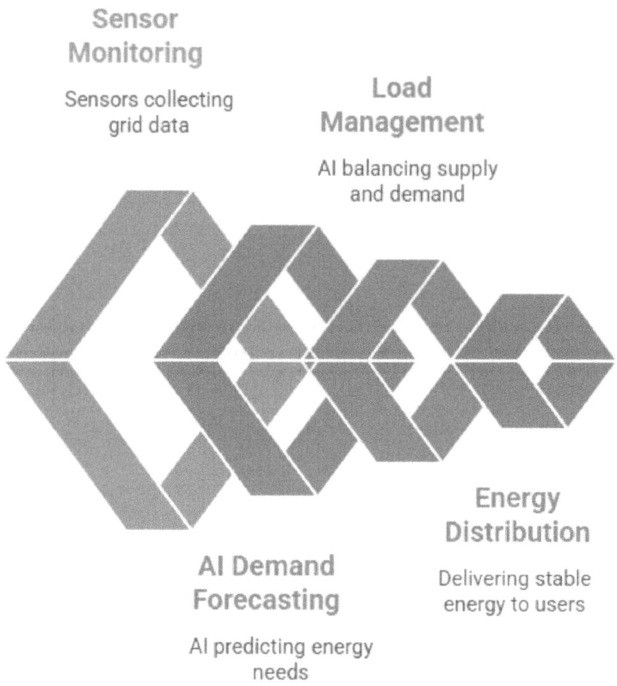

FIGURE 3.6
Smart grid energy management.

3.3.5 Smart Meters: Facilitating Energy Insights in Real Time

Another essential component of this energy-efficient infrastructure is the smart meter. Instead of merely compiling records over time, it displays accurate and thorough usage statistics. As a result, the building manager or the residents themselves can quickly identify resource waste and make the required adjustments. The potential utility of a smart meter for demand-side energy management is among its most important potential benefits. Smart meters that display their energy consumption in real time could influence residents' behaviour. For instance, when energy prices are low, people can turn off appliances during peak hours or perform the least energy-intensive tasks during off-peak hours. The average energy consumption of smart meter users decreased by 12% over the first year of the UK test (Sovacool et al., 2017). In dynamic pricing schemes, where energy costs are contingent on supply and demand, smart meters are also essential. When electricity prices are cheaper during off-peak hours, a building can use smart meters to power high-energy devices like electric car chargers and washing machines. According to the Energy Information Administration, smart meters can reduce energy peak demand by 10–15% because they enable dynamic pricing (Koukouvinos et al., 2025). Thus, this indicates that the strain on the electrical

TABLE 3.5

Smart Grid Capabilities and Maturity Levels across Developed Nations

Country	Smart Grid Maturity	Real-Time Monitoring	Demand Response Integration	Renewable Integration (%)	Outage Reduction (%)	Investment ($B)	National Policy Status
Germany	Advanced	Yes	Fully deployed	48	60	12.5	Mandated
The United States	Mature	Partial	Regionally active	38	45	20.2	Encouraged
Japan	Advanced	Yes	Fully deployed	41	55	10.7	Mandated
China	Developing	Yes	Pilot stage	32	30	15.8	Encouraged
The United Kingdom	Mature	Yes	Fully deployed	42	52	9.6	Mandated
India	Emerging	Partial	Pilot stage	26	22	3.3	Voluntary
Australia	Developing	Yes	Regionally active	39	50	4.1	Encouraged
France	Mature	Yes	Fully deployed	44	58	8.4	Mandated

networks has decreased. Data analytics technologies combine sensors and smart meters to provide energy efficiency and predictive maintenance. Once all the data generated by so many devices is examined, their building management system will be able to identify any discrepancies in their energy use, which typically point to some kind of equipment malfunction or inefficiency. Such a sudden spike in the HVAC system's energy usage would likely be the result of a malfunctioning compressor or a blocked air filter. By identifying and fixing these issues early on, building systems' lifespans are increased and energy waste is decreased. Predictive maintenance also lowers repair expenses and downtime. A case study at one high-rise office building in Singapore found that maintenance expenses may be lowered by 15% and energy expenditures associated with HVAC charges might be lowered by up to 20% using smart meter data for predictive maintenance. This illustrates how using smart technologies for energy management can have both operational and financial advantages (Manivannan et al., 2017).

3.3.6 Automation, Smart Meters, and Sensors: A Trifecta for Maximum Effect

When energy efficiency is approached holistically, sensors, smart meters, and automated energy distribution systems work together. Buildings can develop into intelligent ecosystems where all systems interact with one another in the most energy-efficient manner feasible when these technologies are combined. Through the absorption of sensor occupancy data, smart meter consumption data, and outdoor weather data, a connected building can dynamically modify its energy use throughout the day. In addition to these useful advantages, it ensures sustainability by lowering carbon emissions. For instance, according to 2021 World Green Building Council research, using Industry 4.0 technology has reduced greenhouse gas emissions from buildings and saved energy by up to 30% on average (Asif et al., 2024). These technologies will continue to contribute to global sustainability initiatives if they are used in the building industry. In the heart of energy-efficient construction lies the feedback synergy between physical and digital systems. Figure 3.7 captures this duality through a mirrored framework of a digital twin system. On one side, physical sensor report real-time environmental and operational conditions; on the other side, the virtual model performs simulations, suggesting adjustments for optimization. These recommendations feed back into physical operations such as adjusting ventilation rates based on predicted occupancy.

3.4 Reducing Waste through Predictive Maintenance

A key element of sustainable operations in the building and renewable energy sectors is predictive maintenance. By anticipating probable system failures and taking preventative measures, predictive maintenance lowers

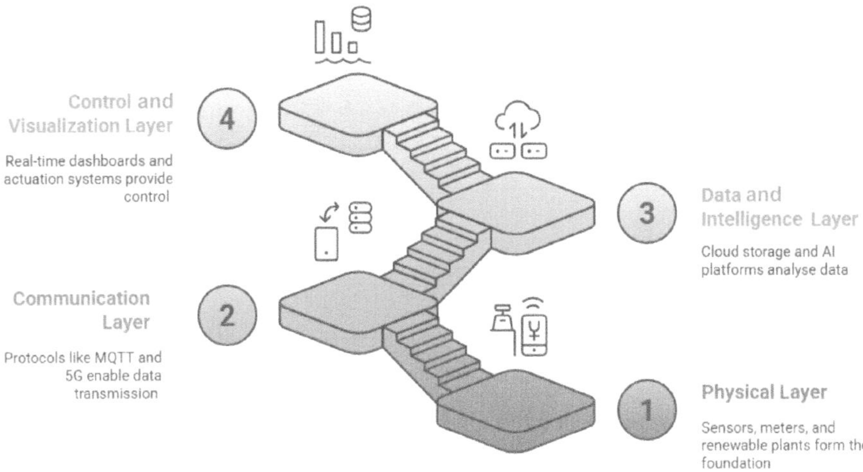

FIGURE 3.7
Industry 4.0 integration in energy systems.

waste, the likelihood of downtime, and operational costs. AI and IoT, two key Industry 4.0 technologies, made the most recent groundbreaking solution possible. This chapter will describe how AI and IoT make predictive maintenance useful in real-world situations, considering its effects in actual case studies that demonstrate cost and waste reduction. IoT plays a foundational role in modern energy management. Table 3.6 details empirical data on how sensor networks contribute to lower maintenance costs, enhanced HVAC responsiveness, and improved power load balancing. The synergy between IoT hardware and cloud-based analytics platforms is especially evident in commercial construction.

3.4.1 AI and IoT's Role in Predicting System Failures

The old method of carrying out maintenance activities for organizations was completely transformed from reactive to predictive by the convergence of AI and IoT. Reactive refers to only responding to occurrences; this type of strategy typically leads to increased expenses, unscheduled downtime, and resource waste. To determine the ideal time to plan maintenance work, predictive maintenance uses real-time data and sophisticated algorithms.

3.4.2 IoT: Information Collection and Real-Time Monitoring

Because they make it simple to monitor systems and equipment continuously, IoT-enabled devices are the foundation of predictive maintenance. Sensor-equipped machines gather data in real time on important parameters, including temperature, vibration, pressure, and energy usage. Cloud-based

TABLE 3.6

Life Cycle Cost Comparison of Traditional versus Smart Renewable Plants

Cost Component	Traditional Wind Plant ($M)	Smart Wind Plant ($M)	Cost Difference (%)	Traditional Solar ($M)	Smart Solar ($M)	Cost Difference (%)	Notes
CAPEX	100	110	+10	80	88	+10	Smart tech adds upfront cost
OPEX (10 years)	45	25	–44	30	18	–40	Smart = fewer failures
Downtime loss	20	5	–75	10	3	–70	Predictive maintenance
Maintenance	15	8	–46.7	9	5	–44	Based on AI use
Energy output loss	12	3	–75	6	1.5	–75	Digital twins optimize ops
Staffing	10	7	–30	6	4.5	–25	Reduced manual checks
ROI (years)	10	6	–40	9	5.5	–39	Faster returns
Net savings	N/A	+31	-	N/A	+26	-	Total cost-benefit

platforms receive the data and use it to look for irregularities or anticipate an impending failure. On a very precise basis, IoT sensors might also be used to monitor wind turbines' blade rotation, gearbox performance, and weather. Operators do this by looking for early indicators of deterioration, like high vibration and odd temperatures, in the data. For instance, a 2021 Deloitte report revealed that wind farms' unexpected downtime was cut by almost 25%, thanks to operational savings from IoT-enabled predictive maintenance (Asif et al., 2024).

THOUGHT BOX

WHO OWNS YOUR ENERGY DATA?

As buildings become digitized and every watt of energy is tracked, the data generated holds immense value. But who owns this data? The building owner? The utility provider? The tenant? In the age of Industry 4.0, data governance will become as critical as energy efficiency. Future legal battles may not be over land, but over data rights tied to smart buildings.

3.4.3 AI: Predictive Algorithms and Advanced Analytics

AI enhances the IoT by sorting through massive amounts of data that human operators could overlook. Using historical and current data, machine learning algorithms generate prediction models that calculate the probability of equipment failure and the suggested corrective measures. A model will get better the more accurate it is in forecasting. While identifying faults to improve the maintenance plan is one of AI's predictive capabilities, it is not the only one. Predictive analytics from the AI systems on RUL components ensure that maintenance is only carried out when necessary, preventing both catastrophic breakdowns and premature replacements. For instance, inverter failures in solar energy systems can be predicted by AI-driven analytics, allowing for prompt repairs and a decrease in energy generation losses. AI-driven predictive maintenance can increase an asset's usable life by 10–15% while lowering costs by 20–30%, per a McKinsey & Company evaluation. Reducing the carbon footprint of construction requires a shift to smart technologies (Pandey et al., 2021). In Table 3.7, we compare CO_2 emissions at each stage of the construction life cycle for legacy and smart approaches. The data reveals that Industry 4.0 integration, particularly during the operational phase, can reduce emissions by over 40% (Javaid et al., 2022).

3.4.4 AI and IoT Integration: The Ecosystem of Predictive Maintenance

Essentially, the secret to integrating AI and IoT into any ecosystem is predictive maintenance. Inferences are made using AI algorithms based on

TABLE 3.7
Energy Efficiency Metrics in AI-Driven Construction Projects

Project Type	Energy Baseline (kWh/m²/year)	AI-optimized (kWh/m²/year)	Savings (%)	Automation Level	AI Model Used	Project Cost ($M)	Post-AI ROI Increase (%)
Office tower	220	145	34.1	High	Neural net	80	27
Hospital	350	260	25.7	Medium	Random forest	100	21
Residential high-rise	180	125	30.6	Medium	SVM	40	19
University campus	310	190	38.7	High	Deep learning	120	32
Smart mall	240	160	33.3	High	LSTM	70	28
Government HQ	300	205	31.7	Medium	Decision tree	95	25
R&D lab	260	180	30.8	High	CNN	60	30
Mixed-use	230	155	32.6	High	Hybrid ensemble	85	26

the unprocessed data collected by IoT sensors. This collaboration guarantees prompt and effective maintenance by means of real-time notifications, automated decision-making, and ongoing monitoring. In smart buildings, IoT sensors will keep an eye on lighting, HVAC, and energy usage. AI systems examine that data to forecast when equipment will malfunction. For instance, the predictive maintenance platform may generate a work order, arrange for a technician, and purchase parts if a malfunction is identified. The connection increases building systems' productivity and dependability while decreasing waste. Real-world case studies demonstrate how much money is saved and downtime is kept to a minimum. The presentations that follow provide examples of how AI and IoT technologies have been used in the real world to minimize downtime, avoid waste, and save money, all of which support the idea of predictive maintenance.

3.4.4.1 Case Study 1: Danish Wind Turbine Maintenance

One Danish powerhouse is setting the norm for renewable energy with a suite of IoT sensors, thanks in part to an AI-powered predictive maintenance system it has implemented for its wind farms. For generators, gearboxes, and blades, respectively, ongoing temperature, vibration, and rotational speed measurements are planned. AI models that recognize wear indications and predict the amount of usable life will be fed these. As a result of the predictive maintenance fixing potentially troublesome issues before any downtime event could happen, there was a 30% decrease in unscheduled downtime. The outcomes were revolutionary: the system helped save unnecessary inspections and repairs, and operational costs decreased by 20% (Javaid et al., 2022). Additionally, it prevented replacements, which usually require a lot of resources, from destroying the environment. The company's sustainability goals were met in part because of this.

3.4.4.2 Case Study 2: Intelligent Building HVAC Systems

For a significant portion of the energy consumption in one of Singapore's numerous tall office towers, predictive maintenance has been implemented. AI algorithms identify irregularities and forecast equipment failure, while IoT sensors track temperature, airflow, and compressor performance. In one instance, the system identified an abrupt increase in energy use as a sign that the compressor was about to fail. A total failure, which would have raised the expense and caused the renters agony, had been avoided, thanks to the urgent maintenance. As a result, within a year, predictive maintenance technology was able to cut downtown HVAC expenses by 40% and energy usage by 15%. It also made renters happy and saved a significant amount of money (Jena et al., 2020).

3.4.4.3 Case Study 3: Predictive Maintenance for Solar Farms

Predictive maintenance was used across all the solar farms owned by a Californian solar energy company to avoid reoccurring problems like inverter failure that could lower energy output. IoT sensors keep an eye on electric flow, temperature, and inverter performance; AI algorithms look for patterns that could point to future malfunctions. The company was able to guarantee uninterrupted electricity generation and cut downtime by 25% by anticipating inverter breakdowns weeks in advance (Hanzel, 2024). Because repairs were completed on schedule and exactly where they were needed, maintenance expenditure decreased by 18%. As energy losses decreased due to better inverter performance, energy efficiency rose by 12% (Bhattacharya, 2024). In this instance, increased operational efficiency and sustainability were two benefits of predictive maintenance. Predictive maintenance powered by AI and IoT is the revolutionary strategy that will eventually allow contemporary companies to cut waste, decrease downtime, and save money. By anticipating system failures, proactive maintenance eliminates the inefficiencies that have long dogged conventional methods. While digitalization brings efficiencies, it also introduces vulnerabilities. Figure 3.8 delineates a three-layer security model specifically tailored for smart construction energy systems. The first layer of device-level security includes measures like secure boot and sensor authentication. The second layer addresses network-level concerns with tools such as firewalls and encrypted channels. The third layer encompasses application-level defences like access control lists and anomaly detection AI. Each segment highlights potential attack vectors and appropriate countermeasures, providing a roadmap for constructing resilient, secure energy infrastructures.

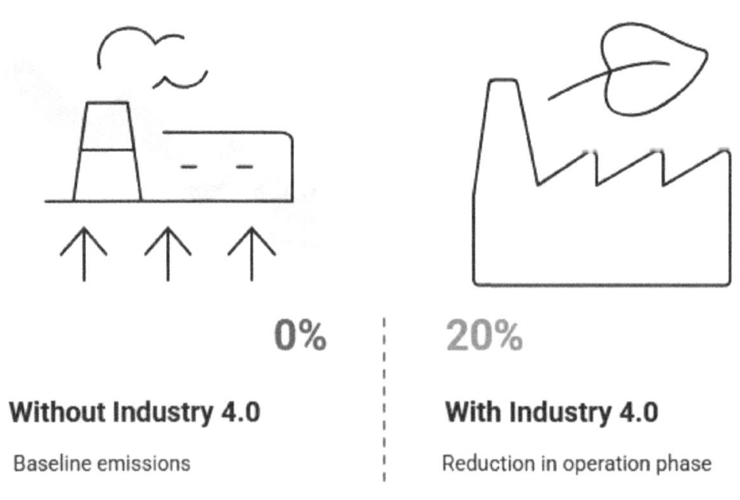

FIGURE 3.8
Carbon emissions with and without Industry 4.0.

3.5 Smart Grids and Decentralized Energy

Decentralized systems and smart grids have now supplanted the one-way and main-controlled grid energy delivery. An efficient, dependable, and sustainable energy supply management system should result from the use of the newest technologies and contemporary advancements in all these new energy sources. When combined with blockchain, the IoT, AI, and cutting-edge digital communication technologies, smart grids enable real-time energy monitoring. It makes dynamic demand–supply management and decentralized trade possible. The function of blockchain in peer-to-peer energy trading is the main topic of this chapter. The management of supply and demand fluctuations by adaptive smart grids is also covered in this section. Put differently, blockchain is a decentralized, irreversible digital record that has recently upended several sectors, including the energy sector.

THOUGHT BOX

FROM STATIC GRIDS TO LIVING ECOSYSTEMS

Traditional power grids operate on top-down control with predictable energy flows. Smart grids, however, function as *dynamic ecosystems*, adjusting in real time to usage patterns, weather, and storage levels. This shift from static to living systems demands not just technology upgrades but also a *philosophical rethinking* of infrastructure as adaptable, reactive, and sometimes unpredictable.

This opens a whole new world of peer-to-peer energy trading and further decentralizes the energy markets by enabling people and small producers, commonly referred to as 'prosumers', to generate, use, and sell energy to others directly, without the need for an intermediary. Large utility companies dominate the energy markets, controlling most of the production, distribution, and billing processes. The efforts to combine diverse and widely distributed renewable energy sources have been primarily centralistic, opaque, slow, and expensive. The open, decentralized market that blockchain-based solutions provide allows energy to be traded in transactions that are tracked on the chain and settled automatically by smart contracts. When certain conditions are met at the time of writing, a smart contract, a set of instructions stored on the blockchain automatically, goes into action. A smart contract controls the exchange of energy in this situation as well; for instance, it transfers surplus solar energy from one subgroup family to another. When specific requirements about the energy supply and price are satisfied, the contract automatically initiates an energy transfer and completes a payment. By doing this, the middleman is eliminated, transaction costs are comparatively minimal, and relationships between multiple parties can be formed without trust. The Brooklyn Microgrid initiative in New York is among the most intriguing instances of blockchain technology being used in the energy trade. The villagers using solar panels were able to sell energy using a blockchain network. Because the local communities could produce and exchange renewable energy, the project lessened their reliance on traditional electrical sources. In a similar vein, Power Ledger, an Australian firm, introduced several blockchain-based energy trading platforms worldwide. Building owners in Thailand were permitted to sell energy as part of a pilot program. This led to increased grid efficiency and energy cost savings of up to 20% (Seeley & Dhakal, 2021). Blockchain-based energy trade does provide a few difficulties. In actuality, the proof-of-work procedures in some blockchain networks consume a lot of energy. The goal of more recent consensus algorithm sets, such proof-of-stake, is to have blockchain operations use nearly no energy. Among the problems that have prevented blockchain use from becoming more widespread include regulatory barriers and a lack of standards. Blockchain technology will become much more integrated into energy markets as it continues to advance and receives increased regulatory approval. This is because energy markets are now more inclusive, transparent, and efficient because of blockchain decentralization. Additionally, it gives prosumers a reasonable return on grid investment while allowing consumers access to locally produced renewable energy sources at reasonable prices. One of the most crucial stages towards a strong and sustainable energy future is the democratization of the energy sector. Predictive maintenance is vital to minimize downtime and extend equipment life. Table 3.8 reports accuracy, false positive rates, and lead time for different AI algorithms (e.g., LSTM, CNN, RF) applied in renewable installations. The high accuracy and early fault detection capabilities emphasize the cost-saving potential of AI in maintenance routines.

3.5.1 Adaptive Smart Grids: Handling Changes in Supply and Demand

Conventional energy systems are very different from smart grids. Smart grids are dynamic, bidirectional networks that facilitate real-time communication and energy exchange between producers, consumers, and grid administrators, in contrast to conventional grids that transport energy on a single path from centralized power plants to customers. On the other hand, adaptive smart grids use cutting-edge technologies to track the fluctuating energy supply and demand in the interest of maximum dependability and efficiency. It is still exceedingly difficult to integrate solar and wind energy, two potentially viable renewable energy sources, into the electrical system with relative ease. In this case, various energy sources produce varying results depending on the weather and time of day. This inherently necessitates the integration of energy storage technologies, IoT devices, AI-powered analytics, and other solutions that will allow for proactive management of smart grid problems. The essential building blocks of smart grids' adaptive operation

TABLE 3.8

Interoperability Standards for Industry 4.0 Technologies in Energy Systems

Standard	Technology Focus	Supported Protocols	Regions Used	Compliance Level	Role in Smart Grids	Device Compatibility	Adoption Rate (%)
IEC 61850	Substation automation	MMS, GOOSE	Global	High	Control and monitoring	High	80
MQTT	IoT messaging	TCP/IP	Global	Medium	Sensor communication	Medium	65
OPC UA	Data interchange	Binary/XML	Europe/Asia	High	Device interfacing	High	72
BACnet	Building automation	UDP/IP	North America	Medium	HVAC integration	Medium	54
Modbus	Legacy systems	RS-485/TCP	Global	Low	Field device control	High	60
IEEE 2030.5	Smart energy profile	Web services	The United States/Canada	High	Grid-edge devices	Medium	48
LoRaWAN	Long-range IoT	Sub-GHz RF	Europe	Medium	Environmental sensors	Low	42
DNP3	SCADA/telemetry	Serial/IP	North America	High	Utility Comms	High	75

are IoT devices. Sensors transmit data and record energy generation and consumption in real time, in addition to smart meters, networked appliances, and grid status. These kinds of systems apply this knowledge and work with it to provide guidance on the best way to distribute energy. It accomplishes this by momentarily modifying the energy supply in response to periods of high demand, such as cutting power to less crucial users while keeping essential users supplied. By using AI and machine learning to sort through a sea of data and extract insightful insights, IoT accelerates that process. From energy consumption and renewable energy production to potential system inefficiencies, grid controllers would now be able to predict it all. Demand–response systems will soon allow smart grids to monitor energy distribution in real time and filter out unbalanced scenarios that cause waste or outages in an entirely AI-driven environment. For instance, the grid balances supply against an unforeseen surge in demand for energy by using stored energy reserves or increasing energy production from other sources. Successfully embedding Industry 4.0 in construction demands more than just technology, it requires coordinated organizational change. Figure 3.9 outlines a structured workflow beginning with vision-setting and stakeholder alignment. It proceeds through phases of infrastructure modernization and workforce reskilling, culminating in deployment and iterative feedback cycles. The diagram clearly marks the responsibilities of key roles like CTOs, system integrators, and regulatory officers. This helps contextualize how people, policy, and platforms must align for sustainable digital transformation.

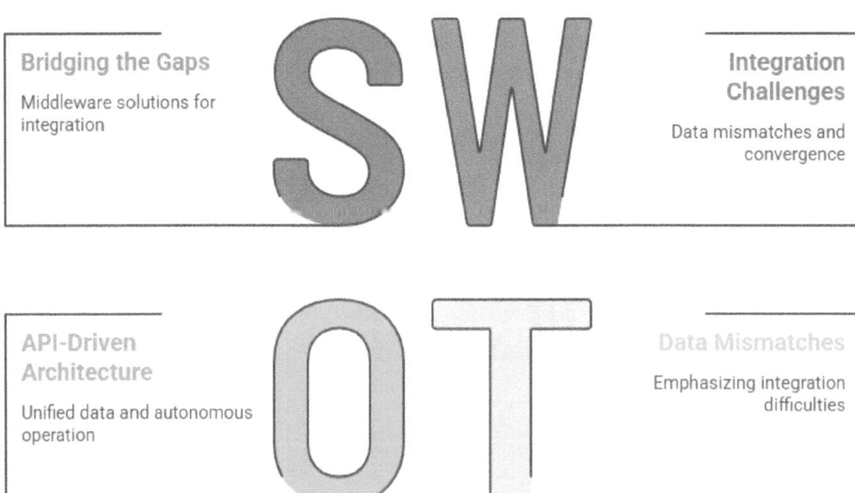

FIGURE 3.9
Legacy versus smart infrastructure.

In fact, In fact, as of 2023, about 52% of the electricity in Germany's grid infrastructure comes from renewable sources, thanks to fully integrated renewable energies and dynamic supply and demand balancing (Kockel et al., 2022). Smart grids represent the backbone of next-gen energy management systems. Table 3.9 compiles key performance indicators from global pilot projects, measuring grid stability, demand–response latency, energy savings, and outage reduction rates.

Through the Smart Grid Investment Grant program, the United States has also provided funding for many projects. In California, a similar project uses IoT and AI to monitor and manage the power supply during months when wildfires occur often. To synthesize the roles of various Industry 4.0 technologies in one cohesive system, Figure 3.10 presents a node-based network that maps inter-system communication. AI nodes connect with IoT modules for real-time data intake, interface with blockchain layers for transaction security, and sync with digital twins for simulation analysis. Each interaction is represented by directional flows annotated with latency, data volume, and dependency scores.

Providing consumers with several opportunities to actively participate in energy management can also increase sustainability and energy independence. However, implementing such a transition to flexible smart grids is quite challenging. Energy storage systems and smart meters are examples of new infrastructure that is so expensive that almost no normal region can possibly afford them. Networked systems raise the possibility of cyberattacks, which can interfere with business operations and steal personal information. To allow third-party innovators to access smart grids, greater regulation will be necessary. Despite the long-term benefits, the issues have made

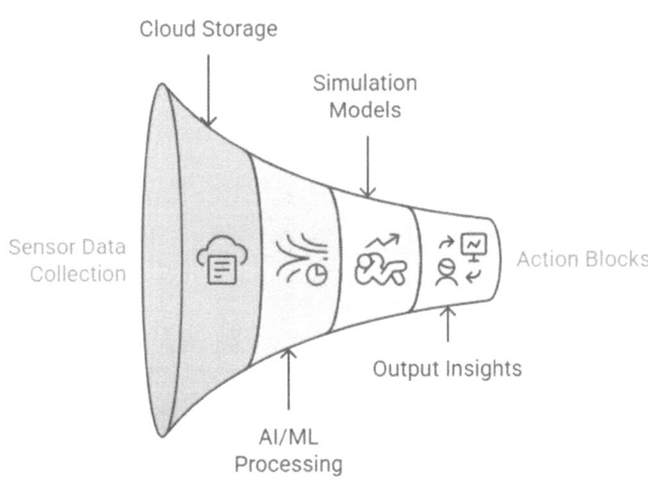

FIGURE 3.10
Data flow in smart infrastructure.

TABLE 3.9
Barriers to Adopting Industry 4.0 in Renewable Energy Infrastructure

Barrier	Category	Impact Severity (1–5)	Example Scenario	Affected Stakeholders	Cost Implication ($M/year)	Mitigation Strategy	Time to Resolve (Years)
Legacy systems	Technical	5	Grid software mismatch	Utilities	3.2	Middleware + upgrade	3
Data silos	Organizational	4	No API access between tools	Project managers	1.5	Unified data standards	2
Cybersecurity risk	Security	5	IoT breach via smart meter	End-users, IT	4.5	End-to-end encryption	2.5
Skill gap	Human resource	4	Lack of AI/IoT training	Workforce	2.0	Upskilling programmes	2
Regulatory Lag	Policy	3	No laws for P2P trading	Governments	N/A	Policy frameworks	4
Interoperability	Technical	4	Sensor can't link to dashboard	Vendors	1.8	Protocol standardization	2
Capital expenditure	Economic	5	High upfront costs	Investors	6.0	PPP or subsidies	5
Cultural resistance	Organizational	3	Manual process preference	Site teams	N/A	Change management	2

flexible smart grids an essential component of the global energy revolution. There will undoubtedly be more chances for adaptive smart grid contributions to sustainability, dependability, and efficiency as digital technologies continue to advance. The most advanced methods for building an effective and sustainable energy infrastructure are today's smart grids and decentralized energy systems. In this case, blockchain technology has improved the transparency and integrity of the energy market between prosumers and consumers. Additionally, innovation and blockchain integration in democratized energy markets enable individuals and communities to take charge of their own energy needs. IoT- and AI-driven smart grid modifications address the problem of incorporating renewable energy sources into the system. By using the optimization principle, they maximize supply and demand losses and provide uninterrupted electricity during demand peaks. Smart grids and distributed energy systems could be a lifeline to transform the green energy supply into a resilient, clean, and far better energy landscape than the one we have today. This is because the green energy supply faces many challenges every day, often because of attention-seeking regulations, cyber risks, and excessive cost-oriented.

3.6 Conclusion

Integrating Industry 4.0 technology with renewable energy systems is likely the most significant and inventive step towards a sustainable and efficient energy future. Regarding urgent concerns like climate change, energy security, and the global increase in demand for electricity, so-called Industry 4.0 technologies such as blockchain, IoT, AI, and advanced analytics will be essential. While some of these advancements enhance energy generation, distribution, and consumption, others further streamline the operation of renewable energy systems. However, how these cutting-edge technologies are repeatedly incorporated into intelligent infrastructure and cooperative ecosystems will determine the direction of renewable energy in the future. Because digital twins are more accurate at modelling, predicting, and optimizing renewable energy projects, they are likely to be employed more frequently soon. To manage wind farms, solar arrays, and smart grids, digital twins are virtual versions of the real systems that may be used to identify possible inefficiencies and implement solutions. Integrating real-time data to maximize energy production, reduce operating costs, and adjust to changing conditions is one of the efficient ways digital twins can assist renewable energy operators. Every day, AI and machine learning can increase the efficiency of renewable energy systems. Soon, demand forecasting, energy storage management, and predictive maintenance will undoubtedly be enhanced by AI in development. Energy waste might be decreased and grid operators' accuracy in energy usage could be improved by employing AI technologies to dynamically balance supply and demand. The most significant change in grid management would occur with the creation of autonomous decision-making algorithms, which would enable renewable energy plants to respond instantly and independently. Peer-to-peer energy trading is another feature that blockchain can provide, making it a powerful advocate for a decentralized energy system or the democratization of the energy industry. It can assist in lowering transaction costs. While the benefits of Industry 4.0 in construction are numerous, challenges remain. Table 3.10 categorizes these into technical, organizational, and regulatory domains. Alongside each issue such as cybersecurity vulnerabilities or legacy system integration, it presents viable mitigation strategies like secure IoT gateways or digital transformation training modules.

The companies will then be able to provide legal compliance documents by enhancing transaction transparency through blockchain technology and enabling prosumers to actively participate in energy trading and renewable energy networks which they may use to monitor and validate their renewable energy credit and meet their environmental targets. IoT-based gadgets and smart sensors are the way of the future for managing and monitoring renewable energy. The energy system's optimized devices will record grid performance and offer essential data on energy generation and consumption. Thus, adding more renewable sources to the current system will be made easier with the integration of IoT and 5G connectivity, which allows for faster and more dependable communication between energy devices. In addition to the field of renewable energy, Industry 4.0 has had a greater positive impact on the building sector. From concept to execution in the value chain of improvement, the use of robots, BIM, and the IoT will define intelligent construction techniques for renewable projects. Digitalized ecosystems with a wide range of stakeholders will be able to complete projects more quickly, more affordably, and with better sustainability results, thanks to Industry 4.0. In addition, energy-efficient structures made possible by renewable technologies, such as rooftop solar panels and smart grids, have been regarded as the most crucial aspect of urban growth. Innovation in the building sector would undoubtedly be pushed against the worldwide goal of a low-carbon transition if advances made feasible by Industry 4.0 were to become a reality. Therefore, as more renewable energy systems become available, more 'Industry 4.0 technology' prospects will offer corresponding advances in accessibility, reliability, and efficiency in energy creation and distribution. Lastly, sustainable development will advance because of governments, businesses, and individuals working together to cut greenhouse gas emissions and switch to cleaner fuels. This will ultimately make it possible for society to embrace renewable energy sources and raise the level of living for coming generations.

TABLE 3.10

Challenges and Mitigation Strategies in Integrating Industry 4.0 with Renewable Energy Systems

Category	Challenge	Description	Impact on System	Mitigation Strategy	Implementation Example	Priority Level	Time Frame to Resolve
Technical	Cybersecurity vulnerabilities	Increased attack surface due to connected IoT and digital platforms	Potential data breaches, system outages	Use of secure communication protocols, edge security, and encryption	IoT firewalls for smart meters in energy grids	High	Short to medium term
Technical	Legacy system integration	Difficulty interfacing old infrastructure with modern tech	Limited data interoperability and system bottlenecks	Middleware platforms and API bridges	SCADA-to-cloud IoT protocol converters	High	Medium term
Technical	Data silos and fragmentation	Isolated data repositories across stakeholders	Inconsistent analytics, slow response	Establishing unified data lakes and open data standards	Common platform for energy suppliers and builders	Medium	Medium term
Organizational	Lack of digital literacy	Limited understanding of advanced digital tools among workforce	Poor system utilization, errors	Upskilling programmes, digital literacy workshops	Industry 4.0 bootcamps for construction site staff	High	Medium to long term
Organizational	Resistance to change	Cultural inertia within firms	Delays in adoption and poor ROI	Change management plans, internal champions	Pilot test cases with proven ROI	Medium	Medium term
Organizational	High capital expenditure	Significant upfront costs of smart tech integration	Budget constraints, delayed implementation	ROI-driven investment models, phased rollouts	Government-funded subsidies for digital retrofits	High	Medium term
Regulatory	Unclear compliance standards	Evolving or inconsistent policy frameworks across regions	Legal risks, cross-border complications	Advocacy for harmonized digital and energy policy frameworks	International green building certification updates	Medium	Long term
Regulatory	Data ownership and privacy concerns	Legal ambiguity over energy and usage data rights	Legal liabilities, loss of user trust	Data governance policies, anonymization, and consent protocols	GDPR-compliant IoT platforms for smart buildings	High	Short to medium term

Chapter Review: Chapter 3 – Renewable Energy and Industry 4.0

Multiple Choice Questions (MCQs)

These questions are designed to reinforce your understanding of how Industry 4.0 technologies integrate with renewable energy systems. Each MCQ is crafted to test both conceptual knowledge and practical implications.

1. Which Industry 4.0 technology enables decentralized peer-to-peer (P2P) energy trading systems?
   A. Internet of Things
   B. Artificial intelligence
   C. Blockchain
   D. SCADA
2. What is a key function of digital twins in energy-efficient smart buildings?
   A. Drafting physical layouts
   B. Real-time simulation and optimization
   C. Managing HVAC sensors manually
   D. Tracking attendance
3. Which method enables predictive maintenance in wind and solar energy plants?
   A. Manual inspection logs
   B. AI algorithms trained on equipment data
   C. Fault-based reporting
   D. Worker-driven alert systems
4. A major cybersecurity concern with Industry 4.0 infrastructure is:
   A. Bandwidth limitations
   B. Staff fatigue
   C. Expanded digital attack surfaces
   D. Smart lighting control
5. Smart grids are primarily used to:
   A. Control lighting systems
   B. Restrict high energy users
   C. Balance real-time energy supply and demand
   D. Increase manual labour

6. How do IoT systems enhance renewable energy efficiency?
   A. Block remote access
   B. Enable real-time monitoring of energy systems
   C. Delay performance reports
   D. Reduce battery backup
7. Interoperability in Industry 4.0 means:
   A. Exclusive software control
   B. Cross-platform system compatibility
   C. One-way automation
   D. No human intervention
8. One of the key organizational challenges in adopting Industry 4.0 is:
   A. Overdependence on robots
   B. Workforce resistance to digital transformation
   C. Excess solar exposure
   D. Faulty weather data
9. Artificial intelligence helps in renewable energy by:
   A. Increasing hardware costs
   B. Running traditional software
   C. Forecasting energy consumption trends
   D. Eliminating solar requirements
10. A widely used wireless communication protocol in IoT-enabled smart energy systems is:
    A. NFC
    B. Bluetooth
    C. LoRaWAN
    D. Ethernet only
11. For a digital twin to function optimally, it must be connected to:
    A. Offline CAD systems
    B. Real-time sensor data
    C. Weather apps
    D. Employee schedules
12. In blockchain-based smart energy systems, smart contracts are used to:
    A. Sign NDAs
    B. Automate compliance
    C. Execute peer-to-peer energy trades
    D. Increase manual input

13. Why are legacy systems problematic in Industry 4.0 upgrades?
    A. They are solar powered.
    B. They lack digital compatibility.
    C. They are cost-efficient.
    D. They rely on LoRaWAN.
14. Which of the following best measures smart grid reliability?
    A. Internet speed
    B. User load
    C. Predictive failure rate
    D. Staff engagement
15. What's an effective solution to overcome data silos in energy infrastructure?
    A. External hard drives
    B. Unified data lakes
    C. PDF-based dashboards
    D. Vendor lock-in

Answer Key

Question No.	Correct Option
1	C
2	B
3	B
4	C
5	C
6	C
7	B
8	B

Question No.	Correct Option
9	C
10	C
11	C
12	C
13	B
14	B
15	B

References

Afzal, M., Khan, A. U., Bilal, M., Ayyub, M. F., & Shoaib, M. (2023). Digital Twins: Facilitating Digital Transformation from and beyond BIM in Construction. 2023 4th International Conference on Data Analytics for Business and Industry (ICDABI).

Ahmed, A., & Khalid, M. (2019). A review on the selected applications of forecasting models in renewable power systems. *Renewable and Sustainable Energy Reviews, 100*, 9–21.

Albreem, M. A., Sheikh, A. M., Bashir, M. J., & El-Saleh, A. A. (2023). Towards green Internet of Things (IoT) for a sustainable future in Gulf Cooperation Council countries: Current practices, challenges and future prospective. *Wireless Networks, 29*(2), 539–567.

Ali, L., Azim, M. I., Peters, J., Ojha, N. B., Bhandari, V., Menon, A., Tiwari, V., Green, J., Muyeen, S., & Simoes, M. (2023). P2P Trading-Enabled Local Energy Market Supplemented with Blockchain Technology: An Australian Case Study. 2023 IEEE Green Technologies Conference (GreenTech).

Alzammar, N. (2023). *Maintenance optimization of offshore wind farms using digital technologies and criticality assessment: Techniques for achieving sustainability*. NTNU.

Asif, M., Naeem, G., & Khalid, M. (2024). Digitalization for sustainable buildings: Technologies, applications, potential, and challenges. *Journal of Cleaner Production, 450*, 141814.

Bhattacharya, V. (2024). *Best operations & maintenance (O&M) practices to be adopted in large-scale grid-connected ground-mounted solar photovoltaic power plants in India*. Book Bazooka Publication.

Borowski, P. F. (2021). Digitization, digital twins, blockchain, and Industry 4.0 as elements of management process in enterprises in the energy sector. *Energies, 14*(7), 1885.

Boumaiza, A., & Maher, K. (2024). Leveraging blockchain technology to enhance transparency and efficiency in carbon trading markets. *International Journal of Electrical Power & Energy Systems, 162*, 110225.

Chatterjee, J., & Dethlefs, N. (2021). Scientometric review of artificial intelligence for operations & maintenance of wind turbines: The past, present and future. *Renewable and Sustainable Energy Reviews, 144*, 111051.

Hanzel, K. (2024). Analysis of financial losses and methods of shutdowns prevention of photovoltaic installations caused by the power grid failure in Poland. *Energies, 17*(4), 946.

Javaid, M., Haleem, A., Singh, R. P., Suman, R., & Gonzalez, E. S. (2022). Understanding the adoption of Industry 4.0 technologies in improving environmental sustainability. *Sustainable Operations and Computers, 3*, 203–217.

Jena, M. C., Mishra, S. K., & Moharana, H. S. (2020). Application of Industry 4.0 to enhance sustainable manufacturing. *Environmental Progress & Sustainable Energy, 39*(1), 13360.

Kabalci, Y., Kabalci, E., Padmanaban, S., Holm-Nielsen, J. B., & Blaabjerg, F. (2019). Internet of Things applications as energy internet in smart grids and smart environments. *Electronics, 8*(9), 972.

Kockel, C., Nolting, L., Priesmann, J., & Praktiknjo, A. (2022). Does renewable electricity supply match with energy demand?: A spatio-temporal analysis for the German case. *Applied Energy, 308*, 118226.

Kong, M., Dong, B., Zhang, R., & O'Neill, Z. (2022). HVAC energy savings, thermal comfort and air quality for occupant-centric control through a side-by-side experimental study. *Applied Energy, 306*, 117987.

Koukouvinos, K. G., Koukouvinos, G. K., Chalkiadakis, P., Kaminaris, S. D., Orfanos, V. A., & Rimpas, D. (2025). Evaluating the performance of smart meters: Insights into energy management, dynamic pricing and consumer behavior. *Applied Sciences, 15*(2), 960.

Lowry, G. (2016). Energy saving claims for lighting controls in commercial buildings. *Energy and Buildings, 133*, 489–497.

Manivannan, M., Najafi, B., & Rinaldi, F. (2017). Machine learning–based short-term prediction of air-conditioning load through smart meter analytics. *Energies, 10*(11), 1905.

Mehmood, M. U., Chun, D., Han, H., Jeon, G., & Chen, K. (2019). A review of the applications of artificial intelligence and big data to buildings for energy-efficiency and a comfortable indoor living environment. *Energy and Buildings, 202*, 109383.

Mohamed, N., Al-Jaroodi, J., & Lazarova-Molnar, S. (2019). Leveraging the capabilities of Industry 4.0 for improving energy efficiency in smart factories. *IEEE Access, 7*, 18008–18020.

Nejat, P., Jomehzadeh, F., Taheri, M. M., Gohari, M., & Majid, M. Z. A. (2015). A global review of energy consumption, CO_2 emissions and policy in the residential sector (with an overview of the top ten CO_2 emitting countries). *Renewable and Sustainable Energy Reviews, 43*, 843–862.

Onwusinkwue, S., Osasona, F., Ahmad, I. A. I., Anyanwu, A. C., Dawodu, S. O., Obi, O. C., & Hamdan, A. (2024). Artificial intelligence (AI) in renewable energy: A review of predictive maintenance and energy optimization. *World Journal of Advanced Research and Reviews, 21*(1), 2487–2499.

Pandey, B. K., Tanikonda, A., Peddinti, S. R., & Katragadda, S. R.. (2021). AI-enabled predictive maintenance strategies for extending the lifespan of legacy systems. *Journal of Science & Technology (JST), 2*(5).

Rojek, I., Mikołajewski, D., Mroziński, A., Macko, M., Bednarek, T., & Tyburek, K. (2025). Internet of Things applications for energy management in buildings using artificial intelligence: A case study. *Energies, 18*(7), 1706.

Saleem, M. U., Shakir, M., Usman, M. R., Bajwa, M. H. T., Shabbir, N., Shams Ghahfarokhi, P., & Daniel, K. (2023). Integrating smart energy management system with Internet of Things and cloud computing for efficient demand side management in smart grids. *Energies, 16*(12), 4835.

Seeley, C. C., & Dhakal, S. (2021). Energy efficiency retrofits in commercial buildings: An environmental, financial, and technical analysis of case studies in Thailand. *Energies, 14*(9), 2571.

Shehabi, A., Hubbard, A., Newkirk, A., Lei, N., Siddik, M. A. B., Holecek, B., Koomey, J., Masanet, E., & Sartor, D. (2024). 2024 United States Data Center Energy Usage Report.

Silva, J. A. A., López, J. C., Guzman, C. P., Arias, N. B., Rider, M. J., & da Silva, L. C. (2023). An IoT-based energy management system for AC microgrids with grid and security constraints. *Applied Energy, 337*, 120904.

Singh, B. J., & Sehgal, R. (2024). Applications of 3D printing, rapid prototyping, and digital manufacturing for low carbon development. In: *Sustainability in smart manufacturing* (pp. 37–60). CRC Press.

Sovacool, B. K., Kivimaa, P., Hielscher, S., & Jenkins, K. (2017). Vulnerability and resistance in the United Kingdom's smart meter transition. *Energy Policy, 109*, 767–781.

Turnbull, A., & Carroll, J. (2021). Cost benefit of implementing advanced monitoring and predictive maintenance strategies for offshore wind farms. *Energies, 14*(16), 4922.

Ukoba, K., Olatunji, K. O., Adeoye, E., Jen, T.-C., & Madyira, D. M. (2024). Optimizing renewable energy systems through artificial intelligence: Review and future prospects. *Energy & Environment, 35*(7), 3833–3879.

4

Robotics in Construction 5.0

4.1 Overview

The next significant development step is established by Construction 5.0, which follows the fundamental framework of Industry 4.0 and Construction 4.0. Industry 4.0 prioritizes automation and intelligent technology integration, whereas Construction 5.0 prioritizes people. In construction, advanced robots and humans collaborate harmoniously to maximize the benefits of both machine performance and human creativity. The primary objective of Construction 5.0 is to address both worker safety and productivity needs. By utilizing technologies like augmented reality, artificial intelligence (AI) tools, and collaborative robots that work alongside humans, our current technology focus improves human work functions. The goal of Construction 5.0 is to strike a balance between direct human involvement in building operations and the advancement of current technologies. The collaboration puts its best effort into developing renewable energy. To use Construction 5.0 principles in the real world, projects with intricate architectural designs and significant construction activity must be chosen with a focus on environmental factors. Given the widespread use of robots in renewable energy projects, bioenergy is an ideal fit for industry objectives. The use of renewable energy sources has long piqued curiosity worldwide.

Concerns about climate change, resource depletion, and energy security have led to large investments in renewable energy sources, including wind, solar, and hydropower in a number of nations. Infrastructure must be scaled up as soon as possible since, according to IRENA, renewable energy capacity will double by 2030. Robotics, with its cutting-edge solutions for process simplification, cost reduction, and safety enhancement, is crucial to achieving these growing demands (Çelik et al., 2022). Solar installation requires a lot of work since hundreds of photovoltaic panels must be positioned precisely. Most traditional procedures are delayed, less effective, and more expensive due to human error. Robotic technology is the answer; with sophisticated sensors driven by AI, the panel installation robots will complete tasks more quickly and precisely than before. For example, "RoboSolar" has developed a robot that can install panels 30% more quickly than a human.

DOI: 10.1201/9781003599272-4

Project schedules are shortened, and staffing costs are reduced. Additionally, drones equipped with thermal imaging sensors are being used more frequently for solar farm maintenance and surveillance. These can promptly spot defective panels or connection problems that could result in less energy being produced. Drone-based inspections helped a large-scale solar farm in California enhance its maintenance efficiency by 40% in one example study. Wind energy is another significant renewable energy source that gains from automation (Yitmen et al., 2024).

A wind turbine's production stage, which includes huge parts like towers, nacelles, and blades, typically takes place in remote or offshore locations. But human input is necessary for cobots to be able to build such massive parts almost flawlessly. Because robots made by firms like ABB Robotics can weld and bolt turbine parts far more quickly and accurately than any human worker, for instance, the structural integrity is ensured. Telepresence robots are primarily used to assist offshore wind farms in overcoming some of the challenges associated with operating in such severe environments offshore and to provide engineers at those offshore locations with a remote view of the assembly procedures. According to data from the Global Wind Energy Council (GWEC), robotics-assisted wind turbine assembly automation has already begun to lower building costs by up to 25% while enabling faster deployment of renewable energy infrastructures. Construction 5.0 is elevated to a whole new level by IoT, AI, and robotics (Musarat et al., 2023). The Internet of Things (IoT) connects these robots via a network of devices, sensors, and control systems that guarantee real-time data interchange and coordination, while AI algorithms enable the robots to operate in dynamic construction scenarios by learning from experience and adapting for optimal performance. For example, IoT-enabled robots with environmental sensors and GPS may build an offshore wind farm (Stephen et al., 2024). These robots could place the turbine components independently, even under the most unpredictable circumstances. Robots driven by AI use topographical data to arrange solar farm panels for optimal solar exposure and energy return. Using AI-driven robots to build wind farms improves turbine alignment and placement, which can result in a 15% boost in energy efficiency (Firoozi et al., 2024). Safety is unquestionably one of the most crucial factors in renewable energy; in the more complex situations like an offshore wind farm or a massive desert solar project, its significance increases significantly.

It can be creative in lowering hazards and safeguarding employees. Wearable robots and exoskeletons are two examples of devices that increase workers' physical capabilities. These boost productivity and lessen the negative consequences of heavy lifting and repetition, which otherwise lead to accidents. Without interfering, the telepresence robots equipped with cameras and communication devices can keep an eye on operations in dangerous locations. It suggests that Shell uses ROVs, conducts safety checks in offshore areas, and creates industry standards for the calibre of work done. According to the Occupational Safety and Health Administration, the use of robotics in building procedures has resulted in a 25% decrease in workplace accidents from renewable energy projects over the past five years (Hassan et al., 2020). In addition to increasing operational efficiency, robotics is essential to renewable energy projects' long-term and financial sustainability.

Additionally, by increasing energy output, decreasing labour expenses, and speeding up building timelines, robots help renewable energy become more competitive with traditional fossil fuels. Robots reduce waste during construction to optimize resource use from a sustainability perspective. Artificially intelligent systems forecast when repairs are necessary to prolong the lifespan of renewable energy sources and avoid expensive malfunctions. One project that involved integrating robotic technologies was a wind farm in Denmark. A 15% cost reduction and a 20% reduction in material waste were among the advantages attained (Mitchell et al., 2022). Global adoption is accelerated by savings like this. No one could dispute that the introduction of robotics raises several significant ethical and labour issues, even though it has been instrumental in making renewable energy feasible. However, upskilling programmes must consider the displacement of more traditional construction occupations to prepare workers for Construction 5.0. However, collaborative robotics will complement human labour, so perhaps here is where the balanced perspective on automation rests. Societies and industries would find themselves investing heavily in job training and education programmes if they used robot armies as the foundation for construction workforces. As part of Germany's "Skills for Industry 5.0," worker development and technology integration are part of it. Construction 5.0 will be the next big thing, combining renewable energy projects with robotics. It developed into robotics to improve human–robot cooperation and equip it with IoT and AI for sustainability and safety, thereby making the planet greener. This compulsive notion of what can be possible with creativity and purpose together is shaped by these, data, and examples. In addition to being efficient and economical, Construction 5.0 will force human inventiveness to guide what is arguably the most crucial choice for the future of global energy (Kolade & Owoseni, 2022).

4.2 Historical Evolution of Robotics in Construction

4.2.1 From Early Automation to Industry 5.0

The development of robots in the construction industry has gone through several stages, from basic mechanization to intricate systems utilizing sector 5.0 components. These stages can only ostensibly illustrate the stages of human progress in achieving an ever-increasing understanding of the automation of anything, from those tiny machines to that enormous one capable of changing life (Huang et al., 2022). Automation was initially conceptualized with the advent of steam engines and cranes at the start of the Industrial Revolution. Even though they made the task easier and more efficient, they still required a lot of labour because they had to do all of their functions by hand. Simple robotic devices began to become widely used in numerous industries by the early 1900s. Early robotic arms, such as the Unimate type, were created in 1961 for manufacturing but later proved helpful in construction for doing hazardous repetitive tasks. Like Industry 4.0, the concept of Construction 4.0 began to gain traction in the late 20th and early 21st centuries (Barata & Kayser, 2023). The transformation of construction over the centuries has been shaped by a continuous interplay of technological innovation and human need. This century's digital advancements ranging from building information modelling (BIM) to drone-based site inspections reflect a leap in both capability and complexity (Yavari & Pilevari, 2020). As the industry shifts from mechanized labour to intelligent assistance, we witness the dawn of Construction 5.0. This transition from mere automation to human-centric collaboration is best contextualized by tracing the evolution of industrial revolutions. Figure 4.1 visually charts this trajectory, illustrating how each phase starting with steam engines in Industry 1.0 through to cyber-physical systems in Industry 4.0 culminates in Industry 5.0's emphasis on synergy between human intuition and robotic precision (Xu et al., 2021).

To appreciate the trajectory of robotics in Construction 5.0, it is crucial to understand the historical milestones that have shaped automation in the industry. The industrial revolutions, from 1.0 to 5.0, mark transformative stages where technology transitioned from steam to electricity, and now to intelligent AI-driven systems. Table 4.1 offers a comparative view of these

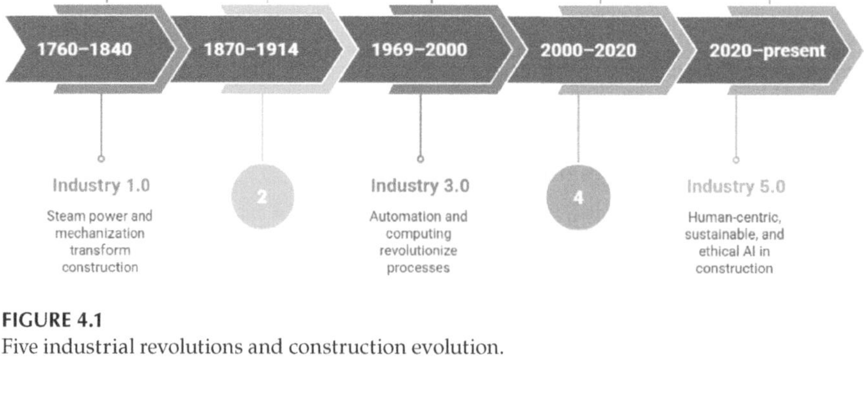

FIGURE 4.1
Five industrial revolutions and construction evolution.

TABLE 4.1

Historical Progression of Automation Technologies in Construction

Industrial Era	Key Technology	Construction Methods	Human Involvement	Primary Power Source	Communication Systems	Safety Standards	Milestone Decade
Industry 1.0	Steam engines	Manual stone/ timber	Very high	Steam	Verbal/symbolic	Non-existent	1780s
Industry 2.0	Electrical grids	Mechanical lifts	High	Electricity	Wired analog systems	Basic PPE	1880s
Industry 3.0	PLCs, robotics	Crane automation	Moderate	Electric + digital	Radio + sensors	ISO-12100 Basic	1960s
Industry 4.0	IoT, BIM, drones	Modular construction	Shared	Renewable + AI	Cloud and wireless	Smart alarms	2010s
Industry 5.0	AI, Co-bots	Robotic teams	Collaborative	Hybrid grids	5G, edge computing	Predictive AI	2020s

stages, mapping the relationship between technology, human involvement, and safety standards across eras. Notably, the evolution towards Industry 5.0 emphasizes hybrid energy systems, AI-based communication, and ethical design focused on human–machine collaboration.

4.2.2 History of Construction Robotic Achievements

Construction robotics is yet another significant technology or application advancement in the construction sector. For example, this would include some significant turning points in the evolution of robotics:

1. **The First Automated Machinery and Automation**
   In the early 1900s, the first automation instruments employed on construction sites were cranes, excavators, and bulldozers. They were not robotic in the contemporary sense, but they did pave the way for automation and significantly decreased the need for human labour. In the middle of the 20th century, basic automation technologies, including conveyor belts and basic robotic systems, started to be developed. These were used in prefabrication procedures to speed up and replicate operations that were similar or identical, like assembling steel framework or concrete panels (Sandler, 1999).

THOUGHT BOX

WHAT IF ROBOTS BECAME CO-WORKERS INSTEAD OF TOOLS?

Imagine a construction site where robots are not just machines executing tasks but active team members. How would workplace culture, communication, and leadership evolve in such a scenario? Would humans lead, follow, or negotiate with AI systems?

2. **Technology for Programmable Machines**
   In the 1960s and 1970s, programmable robotic arms were initially made available. Although their early test applications were successful, it took a long time for them to be used in the building industry. In the 1970s, for instance, Japan's construction sector was only experimenting with automated bricklaying equipment, giving a preview of future developments (Linner & Bock, 2012).
3. **Digital Construction Tools, BIM, and CAD**
   The construction business underwent the most digitization during the 1980s and 1990s. With CAD, it completely changed this area of the design profession. Engineers and architects were able to create precise drawings far more quickly and easily than anyone could have

imagined, thanks to CAD. By combining information from design, building, and operation into a unified digital model, BIM became a huge invention in the early 2000s. Robotics applications began to emerge as these digital technologies advanced. CAD-reading robots started cutting and putting things together. Systems for building automation began to come together (Sepasgozar et al., 2022).

4. **3D Printing and Self-Governing Equipment**

   Early in the 21st century, 3D printing and self-governing construction equipment came together. The necessity for human labour on the job site was increasingly supplanted by self-driving concrete mixers, bulldozers, and excavators. Companies like Komatsu and Caterpillar developed new lines of autonomous heavy machinery that could perform the most difficult tasks with the least amount of human assistance. When it came to home construction, 3D printing was still a relatively new technology. Additionally, it was envisioned that constructions would emerge at amazing speeds, thanks to the sophisticated nozzles of robotic arms that could extrude concrete. The success of Icon, a company that sells 3D printed homes, in building an entire house in a single day lends credence to that (Knips et al., 2014).

5. **Telepresence Robots and Drones**

   Drones and telepresence robots in buildings are examples of other developments. Drone cameras and sensor drones were useful instruments for monitoring developments and assessing the state of construction sites and safety areas. One benefit is that these drones offer a high-resolution, reasonably priced aerial perspective of construction sites. On the other hand, telepresence robots are used to oversee remote building activities. For instance, supervisors can use robots equipped with cameras to explore hazardous regions without physically visiting the locations (Sabet et al., 2021).

6. **Human-Centred Technology and Collaborative Robotics**

Sector Collaborative robots, or cobots as they are more commonly known in the construction sector, have advanced to version 5.0. In contrast to typical robots that operate in a controlled environment, cobots carry out the mission of working alongside people, improving human potential. With the advent of robotic exoskeletons, construction workers can now carry solid masses with significantly less strain and effort. Context-aware AI systems force robots to adjust to their environment, which improves collaboration even more. These technologies enable seamless communication between human workers and robots by interpreting ambient data, human gestures, and voice orders (Boschetti et al., 2022).

The Hadrian X Bricklaying Robot is one of the robotics-related initiatives in the construction industry. Hadrian X, manufactured by the Australian company FBR, can lay 1,000 bricks each hour much more than a human could in an hour. To guarantee accuracy and speed, it makes use of algorithms and 3D design data (Dindorf & Wos, 2024).

- *Boston Dynamics' Spot:* A four-legged, agile robot that gathers information while doing site surveys. The robot is better suited for the construction site because of its agility on challenging terrain.

THOUGHT BOX

COULD TELEPRESENCE REPLACE HUMAN SUPERVISORS ENTIRELY?

Telepresence offers safety and accessibility, but can it fully replace the intuition and quick problem-solving of an experienced on-site supervisor? Should decisions made remotely carry the same weight as those made on the ground?

Workplace safety is one of the most immediate and measurable benefits of robotics. Table 4.2 presents a quantitative comparison of key safety indicators between traditional manual construction and robotic-assisted practices. The integration of wearable robotics, automated surveillance, and predictive monitoring systems significantly reduces the incidence of injuries, fatigue, and life-threatening conditions. Notably, the rate of repetitive strain injuries drops by more than 65% when exoskeletons are deployed, and emergency response systems supported by AI reduce evacuation times by nearly half (Rane et al., 2023).

Semi-automated mason (SAM): Accurate bricklaying lowers labour costs and boosts output. As a result, it is suitable for any building project. The construction industry has benefited from advancements in robotics to an unparalleled degree in terms of cost, efficiency, and safety. The interplay between safety and productivity in modern construction hinges on integrating robots not just as tools, but as co-workers. As highlighted in a McKinsey study, automation can boost productivity by 30% and reduce project timelines by 20% (Anitha et al., 2024). Robots minimize human exposure to dangerous environments, thereby enhancing safety protocols. Rather than operating in silos, human and robotic agents work collaboratively within a shared intelligent space. Figure 4.2 illustrates this multilayered interaction model, where humans use interfaces like exoskeletons, robots handle physical tasks, and both communicate within an AI-enabled framework.

Construction robotics has a long history and a constant need for innovation, from the first automatic tools at the turn of the 20th century to the newest collaborative robots in Industry 5.0. Each new development has expanded the boundaries. Robotics has the potential to revolutionize the construction

TABLE 4.2

Comparative Safety Metrics of Manual versus Robotic Construction

Metric	Manual Construction	Robotic Assisted	% Improvement	Data Source Year	Risk Category	Typical Cause	Mitigation Method
Annual injury rate (per 1,000)	14.3	4.1	71%	2022	Physical trauma	Overexertion	Exoskeleton support
Fatal accidents (per 100k)	9.2	2.7	70.6%	2022	Fall from heights	Lack of railings	Drone and robot monitors
Repetitive strain injuries	33%	11%	66.6%	2023	Musculoskeletal	Manual drilling tasks	AI-synced drills
Hearing loss reports	18%	6%	66.6%	2023	Noise pollution	Machinery operation	Sound-dampening bots
On-site fire incidents	5.1 per site/year	1.2	76.4%	2021	Electrical faults	Overload or short circuit	Predictive IoT alerts
Safety audit failures	21%	6%	71.4%	2022	Compliance	Improper logs	Blockchain-based records
Emergency evacuation time (min)	9.3	4.5	51.6%	2023	Disaster response	Delay in alerts	Smart helmet notifications
Heatstroke incidents (Summer)	12.6	3.9	69%	2023	Environmental	Lack of cooling	Climate-aware uniforms

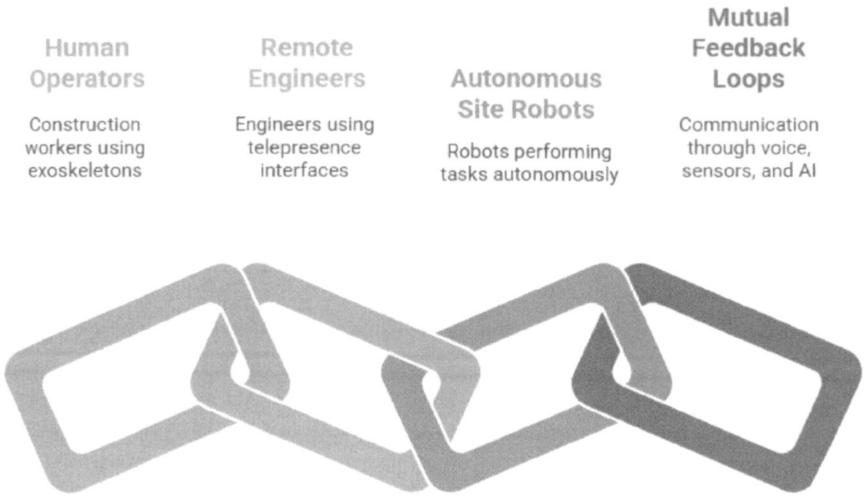

FIGURE 4.2
Collaborative construction ecosystem.

business by making it safer, more efficient, and more ecologically sensitive than it has ever been.

4.2.3 Important Robotics Uses in Renewable Energy Initiatives

Renewable energy initiatives utilizing robotics promise a previously unheard-of breakthrough in creativity, security, and production. Indeed, to meet the obstacles posed by remote locations, dangerous environments, and unpredictable site circumstances during construction or maintenance, renewable energy infrastructures desperately need the newest, most cutting-edge technologies. Among the pioneers of this revolution are applications such as telepresence robotics, exoskeletons, and context-aware robots (Hassan et al., 2020).

4.3 Remote Operations in Dangerous Environments with Telepresence Robotics

Engineers and workers may engage with dangerous environments and typical occurrence in renewable energy projects in a completely different way, thanks to telepresence robotics. Indeed, by enabling remote monitoring and control, these robots enable humans to do vital tasks without having to be physically present in such dangerous situations. For instance, despite being a vital component of renewable energy, offshore wind farms provide considerably higher danger because of their position in choppy waters with unfavourable weather. Wind turbine inspections are conducted by robots equipped with cameras, sensors, and communication devices; operators are not physically present outside of an onshore control room (Walker et al., 2021). Nowhere is this integration more evident than in offshore wind installations, where remotely operated underwater vehicles (ROVs) inspect submerged structures with unparalleled precision. Previously life-threatening tasks were now completed safely through robotic autonomy. But for robots to adapt in real time, context awareness is critical. Figure 4.3 outlines the architecture of such robots detailing layers from sensor inputs and AI-based contextual interpretation to adaptive response systems. This ensures dynamic responsiveness to complex environments.

THOUGHT BOX

ARE EXOSKELETONS EMPOWERING OR ENSLAVING WORKERS?

While exoskeletons reduce fatigue and injury, could their use also pressure workers into higher output without breaks? How should labour laws adapt to ensure human dignity and fair workload distribution?

Telepresence robots have shown themselves to be quite helpful, particularly for solar farms that handle large installations of photovoltaic panels. Using thermal imaging cameras and sensors, the robot would identify hotspots or

FIGURE 4.3
Robotic system process.

broken panels, speeding up repairs. Telepresence robot deployment led to a 35% boost in inspection efficiency and overall favourable operating performance, according to a case study conducted in Nevada. Telepresence robotics improves efficiency and safety while lowering expenses. By eliminating downtown and the need for on-site staff, robots increase the economic feasibility of renewable energy projects. This kind of technology is a great example of how robotics may be used to address the dual problems of safety and scalability in renewable energy (Duncan et al., 2018).

4.3.1 Exoskeletons: Promoting Employee Safety and Productivity

Firstly, wearable robots' exoskeletons are one of the key areas of renewable energy that robotics innovation addresses. This will increase people's potential, mostly by making many occupations easier than they were previously thought to be. In addition to improving worker safety, they lessen weariness brought on by heavy lifting or repeated work that is frequently encountered during large-scale projects like solar panel installation or wind turbine installations. For example, wind turbine nacelle and blade assembly involves hundreds of pieces, and handling them carefully can lead to musculoskeletal injury. Companies like Sarcos Robotics and Ekso Bionics are developing supportive robotic exoskeletons that equally distribute weight to help prevent workers from becoming fatigued and to support their arms, legs, and back. Workers will be able to operate heavy machines more precisely and safely, thanks to these robotic exoskeletons, which will also increase production. Understanding the operational capacity of robotic tools is essential for planning effective construction strategies. Table 4.3 details the technical specifications of devices commonly deployed in renewable projects, such as cobot arms, exoskeletons, and drone surveyors. The table underscores how these technologies are optimized for task-specific performance, including payload capabilities, power duration, and human–robot interface methods. For example, AI-synced crane bots are capable of lifting over 3,000 kg with enhanced safety and precision during wind turbine assembly (Lai et al., 2021).

For solar project workers who install photovoltaic panels, exoskeletons are more advantageous. Employees can attach panels using this equipment while keeping good posture and avoiding constant tension. In the case study of a German solar energy company, this led to a 25% improvement in installation speed and a 40% drop in worker weariness (Simas & Pacca, 2014). In addition, exoskeletons contribute to workforce retention and the expanding labour diversity. These have made heavy chores easier to undertake and allowed individuals who are physically incapable or elderly to work efficiently in renewable energy-related operations. In line with human-centred automation, technologies like exoskeletons extend the capacity of human workers while telepresence robots enable remote management of high-risk zones (Su et al., 2023). Such interventions can significantly extend worker productivity and improve task accuracy. Figure 4.4 depicts how telepresence robots enable

TABLE 4.3

Technical Specifications of Robotic Devices Used in Renewable Projects

Device Type	Weight (kg)	Operating Time (h)	Control Method	Max Payload (kg)	Key Feature	Typical Application	Recharge Time (h)
Exoskeleton (full)	12.5	6	Motion-sync	80	Load redistribution	Panel mounting	1.5
Drone surveyor	2.1	4	Tablet/gesture	5	Thermal and LiDAR combo	Site mapping	2.0
Telepresence rover	18	10	VR/Joystick	0 (non-load)	360° Camera + Mic	Remote inspections	3.0
Co-Bot Arm (6 DOF)	28	8	AI predictive	35	Haptic feedback	Bolt fastening	2.5
Automated crane bot	1,200	12	AI Sync w/ BIM	3,000	Precision lift assist	Turbine blade handling	4.0

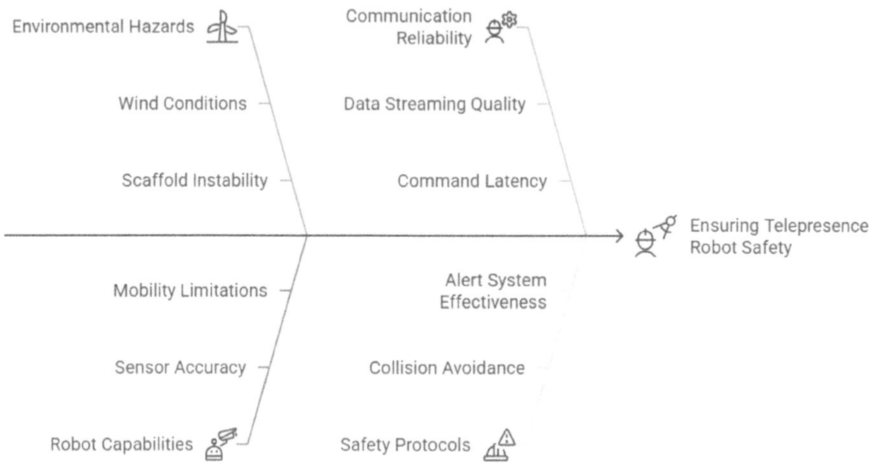

FIGURE 4.4
Enhancing telepresence robot safety in high-risk environment.

real-time surveillance and remote intervention, offering engineers the ability to operate in hostile or elevated environments without physical risk.

4.3.2 Context-Aware Robots: Flexible Robots for Changing Building Environments

These are extremely dynamic and frequently rather surprising, ranging from offshore wind farms to enormous stretches of open fields of solar panels in far-flung deserts. Context-aware robots with advanced sensors and AI will be the primary tools used in these difficulties. The robots would also become extremely perceptive, conscious, and sensitive to even the smallest nuances in the production of renewable energy. In wind energy projects, context-aware robots are now able to assemble and maintain turbines. Turbine parts are precisely located and aligned using autonomous robotic arms and an AI-powered vision system. To ensure uninterrupted work, these robotic arms adapt to variations in wind and weather conditions. Context-aware systems also find widespread application in renewable installations. On solar farms and wind sites, GPS-guided autonomous vehicles and AI-enhanced drones manage panel placements, terrain navigation, and even damage detection. These machines are instrumental in maintaining construction quality and efficiency. Figure 4.5 showcases an integrated workflow, from drone-based surveying and robotic transport to automated component installation and inspection, demonstrating robotics' pivotal role across the project life cycle (Stipancic et al., 2016).

Perhaps the most promising aspect of these context-aware robots is their capacity to operate effectively alongside human employees. Robots at work sites automatically recognize human behavioural clues, spoken commands,

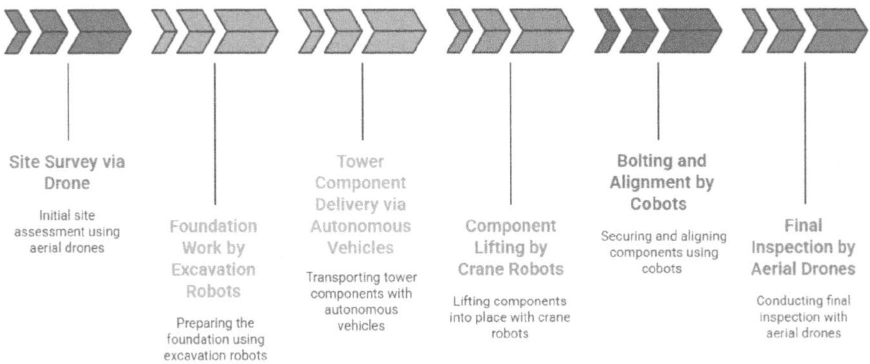

FIGURE 4.5
Robotic wind turbine assembly process.

and gestures. Workers on the North Sea wind farm project, for instance, lifted and installed turbine blades using context-aware robots that vocally instructed workers on the blade's precise location. It consequently became more efficient overall and decreased installation errors. Context-aware robotics has various environmental and financial advantages. Robots assist renewable energy projects in achieving sustainability goals by optimizing resources and reducing material waste. Resilience and reliability are linked to flexibility, which guarantees that initiatives will continue even in the face of unforeseen obstacles. Exoskeletons, context-aware robots, and telepresence robotics are some of the most prominent robotics used in renewable energy initiatives to date. Telepresence robots, which increase security and productivity while working on-site in hazardous environments; exoskeletons, which improve worker performance and well-being in the most taxing of tasks; and context-aware robots, which can adjust to changing work conditions for precise and effective completion of many building-related tasks, are just a few of the innumerable revolutionary technological advancements that will impact the future of sustainability. They would tackle issues in actual renewable energy projects while also advancing the overarching goals of Industry 5.0 and Construction 5.0 in terms of sustainability, economic growth, and human–machine collaboration. The robotics industry will continue to be crucial to the development of clean energy solutions in a more efficient, secure, and pervasive way as the renewable energy sector grows. These cutting-edge technologies accelerate the building industry's transition to more intelligent and environment-friendly futures. Robotic construction practices not only enhance efficiency but also minimize environmental harm. Table 4.4 compares several ecological impact metrics between manual and robotic operations, revealing significant reductions in carbon emissions, water consumption, dust generation, and non-recyclable waste. These results are especially relevant in the context of global climate targets,

TABLE 4.4

Economic Analysis of Robotic versus Manual Construction in Renewable Projects (Hassan et al., 2020)

Project Type	Manual Cost/MW (US$)	Robotic Cost/MW (US$)	ROI (%)	Payback Period (Months)	Labour Savings (%)	Waste Reduction (%)	Uptime Increase (%)
Wind farm (onshore)	1.65M	1.22M	21.3	42	37	29	15
Solar farm (utility)	0.98M	0.79M	25.7	33	28	35	12
Offshore wind	3.1M	2.48M	18.9	56	38	31	22
Smart building	2.3M	1.71M	24.1	39	34	25	18
Green retrofit	1.1M	0.83M	27.2	31	41	40	17

as robotic solutions align with low-impact construction methodologies and reduce site-based environmental degradation.

4.4 Synergies between AI, IoT, and Robotics

New design and deployment choices for renewable energy projects are made possible by the convergence of robotics, AI, and the IoT. Machines and systems interact, communicate, and work together to maximize production, accuracy, and efficiency in this dynamic environment. Using collaborative robotics on large projects up to smart installation precision based on renewable energy is an example of a pioneering move. AI, IoT, and robotics may have the biggest impact on renewable energy by improving the accuracy and efficiency of installations. Careful planning and execution are necessary due to the complexity of installing renewable power plants, particularly large-scale wind and solar farms. Any installation error puts the plants at risk of many delays or extremely low energy output. By enabling automated execution and data-driven decision-making, the implementation of such technology reduces such risks (Ness et al., 2023).

4.4.1 AI-Powered Accuracy

AI algorithms play a major role in the creation of any data used to plan and then carry out each renewable energy producing project. For instance, while building solar farms, AI algorithms analyse terrain, satellite imagery, and

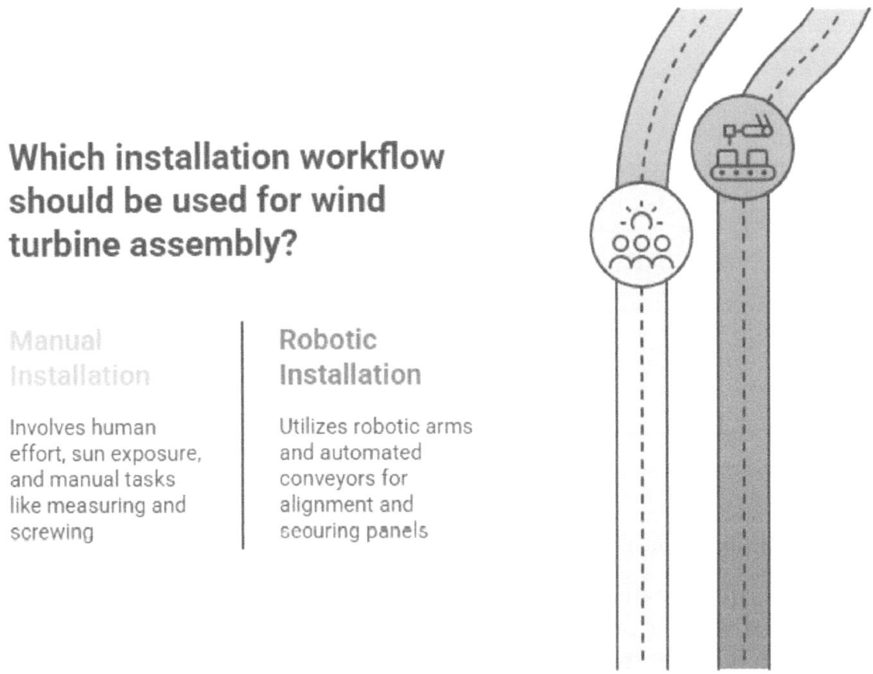

FIGURE 4.6
Workflow for wind turbine assembly.

weather patterns to determine the optimal location for each photovoltaic panel to maximize energy production by maximizing sun exposure. Similar comparisons between wind flow, topographical features, and structural data are used by AI systems to find the optimal turbine position for onshore wind energy projects (Ness et al., 2023). Comparing traditional manual methods with robotic solutions reveals compelling efficiencies. While human teams can suffer from fatigue and variable output, robotic systems consistently deliver high precision and safety. Figure 4.6 presents a comparative performance matrix between manual and robotic workflows, emphasizing gains in time efficiency, error reduction, and worker safety during solar field deployments.

IoT for Real-Time Monitoring and Feedback: Real-time monitoring and reporting from IoT offers an additional degree of accuracy. As construction sites become digitized ecosystems, data plays a transformative role. At a Danish wind farm, IoT integration cut installation errors by 20% and improved scheduling by another 20% (Adekanbi, 2021). This real-time optimization is enabled by continuous data loops. Figure 4.7 illustrates an IoT–AI feedback model where sensor data is processed to adapt robotic behaviour, ensuring real-time responsiveness to site conditions.

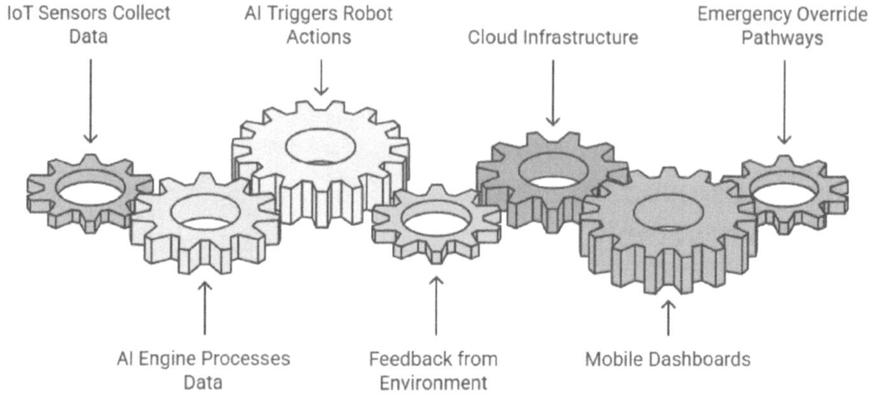

FIGURE 4.7
Smart construction ecosystem control loop.

4.4.2 Robots for Automated Performance

The physical embodiment of precision, the robot performs tasks with an unmatched degree of accuracy. AI-enabled and IoT-capable autonomous robots can do repetitive and unpleasant activities that require human assistance. Examples include solar panel installation robotic arms that use real-time information from IoT sensors to precisely guide and secure panels at the millimetre level. Underwater robots equipped with AI algorithms can perform extremely difficult jobs like anchoring turbine foundations and installing subsea cables in offshore wind farms. The robots are designed to remain steady and dependable in dynamic situations, such as shifting seabeds or oscillating currents. Under such circumstances, the use of robotics improves task completion precision and safety without putting people in danger. Many teams, technologies, and systems are involved in the most significant renewable energy projects. The new collaborative robots, dubbed the next frontier, are made for a machine–man interaction that combines human intuition with the accuracy of machines, as opposed to conventional robots that are made to work alone. Here are some examples of how cobots in large-scale renewable energy projects increase human productivity. The wind turbines, which are massive and have parts for mounting nacelles and blades, have sensors installed and AI algorithms that work in perfect harmony to decode spoken commands and the movements of nearby people. Cobots also put together 8 MW turbines on one of the biggest wind farm projects in the North Sea, which is another intriguing fact. By guaranteeing correct blade alignment and attachment, the robots assist the human workers, reducing assembly time by up to 25% without sacrificing precision or quality. Application is one of the most crucial elements of solar energy installation. The employees of these massive solar farms must oversee tens of thousands of photovoltaic panels. Cobots use automated grippers and vision systems to mount panels more quickly and precisely than human operators. Furthermore, these robots may function flawlessly for the duration of a project since they can easily adapt to changing site conditions. Robotic construction practices not only enhance efficiency but also minimize environmental harm. Table 4.5 compares several ecological impact metrics between manual and robotic operations, revealing significant reductions in carbon emissions, water consumption, dust generation, and non-recyclable waste. These results are especially relevant in the context of global climate targets, as robotic solutions align with low-impact construction methodologies and reduce site-based environmental degradation (Liu & Althoff, 2020).

When cobots and autonomous machinery collaborate, large tasks can be finished much more quickly. For instance, at wind farm installation, the cobots will precisely locate and install each component as soon as the autonomous vehicles deliver the turbine parts to the assembly location. This machine-to-machine cooperation expedites project deadlines and eliminates logistical bottlenecks. Cobots and self-propelled vehicles use the detailed maps produced by very sophisticated AI-driven drones with cameras to distribute panels as effectively as possible. Just-in-time completion with few mistakes would be the outcome. It would include providing cobots with AI and IoT capabilities the intelligence to make some well-informed decisions and adjust under specific circumstances. AI systems will eventually assist

TABLE 4.5

Environmental Impact Metrics: Manual versus Robotic Construction Sites

Impact Category	Manual Method Value	Robotic-Assisted Value	Unit	% Improvement	Primary Source of Reduction	Region Data (Average)
CO_2 emissions	7.1	4.6	Tonne/site	35.2	Less onsite equipment	EU-27
Water usage	22.4	13.9	kL/day	37.9	Dry robotic processing	The United States
Noise pollution	87	63	dB(A)	27.5	Quieter machines	Global Avg
Dust/PM emissions	112	67	$\mu g/m^3$	40.1	Enclosed robotic cutting	China
Material waste	18.3	10.4	% of Input	43.2	Optimized AI usage	Australia
Non-recyclable waste	6.9	3.7	% of Output	46.3	3D printed elements	Scandinavia
Energy efficiency	78	91	% Efficiency	16.7	Grid-synced systems	Germany
Biodiversity disruption	High	Low	Qualitative	N/A	Drone-surveyed zoning	Canada

them in learning, and productivity will rise because of such cooperation with people or other robots. However, as part of the system, the IoT-enabled cobots will exchange data with nearby sensors and control systems. AES has deployed AI-powered solar robots (Maximo) at the Bellefield project in Kern County, California, and Terabase Energy has introduced its Terafab automated "field factory" system, both using robotics, digital twins, and IoT feedback to improve installation speed, safety, and productivity in utility-scale solar projects. (Factor This, 2025; PV Magazine, 2023). Integration could cut the installation time by up to 30% as compared to the conventional approach. The integration of AI and the IoT has dramatically enhanced the responsiveness and intelligence of robotic systems in construction. Table 4.6 highlights different AI–IoT applications, from fall detection to route planning, alongside their reaction speeds, automation levels, and feedback types (Ukoba et al., 2024). These systems enable robots to make real-time decisions, improving task precision and safety while reducing their dependency on constant human supervision.

This results in several advantages for renewable energy projects, including cost reduction through automation, more accurate operations with fewer errors, and quicker project schedules; safety since robots reduce human exposure to hazardous jobs; and efficiency in integrating such technology without delay and full utilization of resources. This has made it possible for these technologies to support the rapid construction of renewable energy projects to satisfy demand installation and collaboration that minimizes waste and aligns with the implementation of renewable energy systems for environmental sustainability. The convergence of robotics, AI, and IoT offers more than operational benefits, it supports long-term sustainability. Automated systems minimize material waste, optimize energy consumption, and reduce emissions. Figure 4.8 uses a Venn diagram to represent energy efficiency, material optimization, and carbon reduction intersect, reflecting how Construction 5.0 meets global climate and ESG targets.

4.5 Economic and Sustainability Impacts

The construction industry has responded by integrating robotics, AI, and the IoT into renewable energy projects, which have transformed the economy and environment. Advanced technologies are speeding up the implementation of renewable energy solutions by automating processes and lowering prices. This shift to clean energy will have a significant impact on communities, governments, and corporations alike because it is both fiscally and environmentally feasible. In summary, automation is being utilized more and more to lower the cost of renewable energy projects. Conventional construction methods for wind and solar systems result in lengthy project timelines,

TABLE 4.6

AI and IoT Capabilities Integrated in Construction Robots

Capability Area	Example Application	Data Type Used	Sensor Count	AI Model Type	Reaction Time (ms)	Automation Level	Feedback Loop
Safety monitoring	Fall detection	Accelerometer + video	6	CNN + Edge AI	200	Level 3	Real-time
Task optimization	Concrete pouring schedule	Time + weather + soil	8	Reinforcement learning	500	Level 4	Batch sync
Resource forecasting	Material use predictions	IoT flow sensors	4	Predictive regression	1,200	Level 3	Daily
Fault detection	Motor vibration analysis	Accelerometer + audio	7	Anomaly detection	300	Level 2	Real-time
Route planning	Site navigation	GPS + LiDAR	5	SLAM + Heuristic AI	100	Level 5	Instantaneous

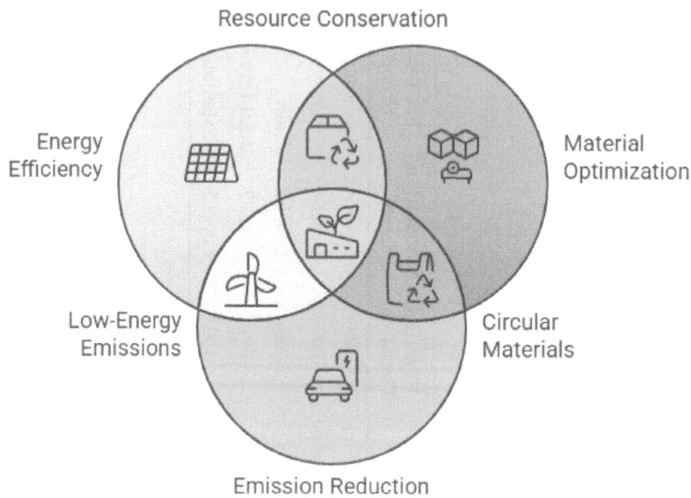

FIGURE 4.8
Synergy for sustainable construction.

significant material waste, labour-intensive procedures, and time-consuming activities. Automation, made possible by robotics and AI, minimizes waste by maximizing resources and expediting procedures. One of the primary ways automations lower costs is by increasing labour productivity. Better sensors, processors, and most significantly, algorithms have made it possible for robots to execute repetitive, more complex jobs with previously unheard-of speed and accuracy. Installing the photovoltaic panels used in solar projects is one example. One farm in Texas attests to the fact that using a robot installation method increases project managers by 40% because it requires less installation manpower.

THOUGHT BOX

WILL GREEN ROBOTICS JUSTIFY HIGH INITIAL COSTS?

Robotic systems for renewable energy construction are expensive to implement. Can the long-term environmental and operational savings outweigh the initial investment? What financial models support this shift?

In this sense, mechanical robots carry out crucial construction-related activities, such as installing wire, erecting turbine blades, and securing the base. In addition to lowering errors, rework, and lengthier turnaround times, productivity also directly lowers labour expenses because of their ability to work continuously, which ensures dependability on time and avoids delays.

Additionally, automation lowers waste and raw material processing costs. Even the smallest errors result in significant material losses for large-scale renewable energy installations. These robots utilize companion IoT sensors and AI-based decision-making systems. Once a turbine is installed, robot arms will install pieces millimetre by millimetre to ensure proper utilization of these resources. It prevents the introduction of modification processes that could jeopardize the robot's structural integrity. Devices with IoT capabilities track the production and consumption of solar energy to position each panel for optimal performance. Process automation reduced material waste by 15% in a case study on an Australian solar farm, which significantly reduced costs. Automation offers further long-term financial advantages outside of the building sector in addition to increased operational and maintenance efficiency. IoT sensors integrated into renewable energy infrastructure support predictive maintenance by continuously monitoring equipment performance. Sensors in wind turbines can identify early wear and tear issues, allowing operators to take corrective action before they become expensive breakdowns. As the nature of construction work evolves, so must the skill sets of human workers. Table 4.7 outlines the transition from traditional manual skills to the modern competencies needed to operate and collaborate with robotic systems. This matrix includes required training time, complexity, and certification needs across various domains such as mechanical installation, surveying, and safety compliance. The data shows that while new skills may require higher initial training investment, retention and productivity outcomes justify the effort.

That is comparable to how AI-driven imaging drones precisely detect flaws in solar farms and wind turbines. Human checks, which can be tedious and time-consuming, are no longer necessary with such a procedure. Operators of renewable energy benefit greatly from this process of automation, which

TABLE 4.7

Worker Skill Evolution Matrix in Robotic Construction

Skill Domain	Traditional Skill	Required New Skill	Learning Curve (1–5)	Average Training Time (h)	Training Mode	Certification Required	Retention Rate (%)
Mechanical	Wrenching	Co-bot programming	4	48	VR Sim + workshop	Yes	84
Surveying	Manual levelling	Drone mapping	3	36	AR modules	Yes	78
Safety compliance	Visual checklists	Digital sensor monitoring	2	24	eLearning	No	91
Installation	Manual fastening	Exoskeleton Integration	3	30	Field simulation	Optional	88
Project planning	Gantt charts	BIM-driven AI scheduling	5	64	AI Labs + remote tutor	Yes	72

also preserves dependability. One of the main motivations behind renewable energy technology is automation. The availability and scalability of renewable energy sources are pushed by advanced technology with a variety of logistical, cost-saving, and efficiency-boosting options. Projects using renewable energy should be scalable to reduce carbon emissions and supply the world's energy. Automation will expedite infrastructure development, enabling governments and businesses to expand renewable initiatives. The time required to develop the construction schedules for the solar farms was actually cut by 30%, thanks to the self-governing robots. The developers may complete the work far sooner than they had anticipated and save a significant amount of time if they are able to complete activities like wiring and site preparation at previously unheard-of speeds and accuracy. When used in areas like the European Union, which aims to achieve 42.5% renewable energy by 2030, the strategy works better (Gajdzik et al., 2024). Additionally, automation makes it simpler to build offshore wind farms. Wind farms are inherently challenging because of their challenging and typically inaccessible on-site surroundings. Self-governing underwater robots and AI-powered logistics systems expedite the installation of subsea cables and turbines, resulting in a quicker project completion time. Wind farms that are developed more quickly have the potential to produce more energy and take less time for investors to recover their investment. For renewable energy projects, automation lowers entrance barriers. The high prices of labour and construction materials have long prevented small- and medium-sized enterprises and developing countries from implementing renewable energy technology due to significant investment barriers. As a result, automation lessens these challenges and provides affordable solutions that can enable stakeholders to take part in the transition to renewable energy. Adopting these technologies also requires reimagining the workforce. New roles such as drone technicians and AI analysts are replacing traditional positions. Upskilling pathways become essential. Figure 4.9 maps the transformation from traditional construction roles to advanced tech-driven careers, emphasizing the training milestones required at each stage.

Deployment of robotics in solar panel installation offers measurable improvements in speed, accuracy, and operational uptime. Table 4.8 provides a side-by-side comparison of robotic-assisted and manual methods, showing substantial reductions in defect rates and downtime, with increased energy output. These improvements not only lower operational costs but also contribute to faster scaling of solar infrastructure to meet rising energy demands.

The US Inflation Reduction Act, which includes several sections on tax credits and other incentives related to projects combining renewable energy sources with advanced robotics and IoT technology, serves as an example in this regard. It encourages investment and consumption expansion by establishing an ideal environment. The expenses and procedures involved in constructing renewable energy infrastructure are increasing since automation is essential to completing the project in the period needed to help the

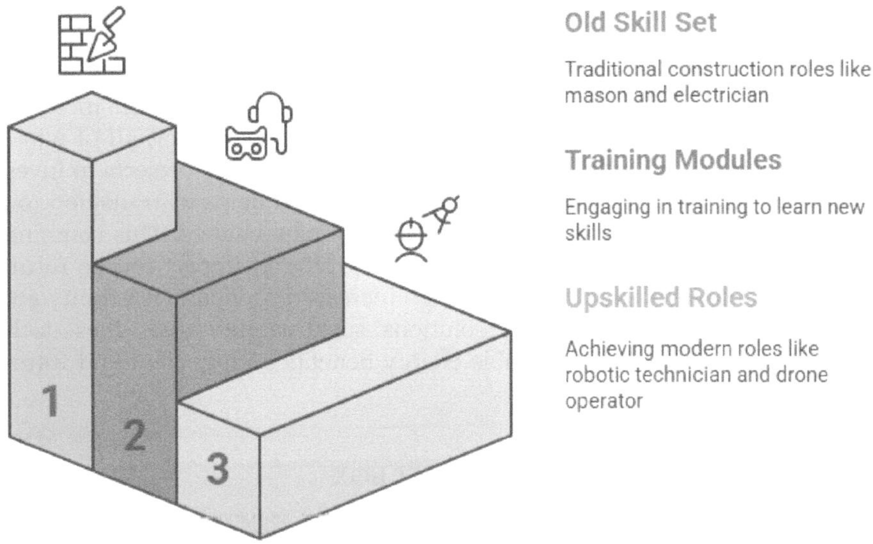

FIGURE 4.9
Transition to modern construction roles.

world reach its climate goals. In this sense, the advantages of automation for the economy and sustainability are interchangeable. Utilizing resources as efficiently as possible in renewable systems reduces waste and, in turn, the carbon footprint. For instance, less emissions are produced during the manufacture of additional materials and their transportation to the construction site when they are handled non-destructively. Drones and self-driving cars lessen reliance on fossil fuel–powered equipment, which further reduces emissions. Through employee upskilling and the creation of possibilities in technology-oriented jobs, it fosters sustainability. The future lies in automation, which frees up workers to do higher-value tasks like system design, data analysis, and equipment optimization. In doing so, it enhances job satisfaction and guarantees a consistent and reliable labour supply for the renewable energy industry. Over time, it also lowers automation costs, making renewable energy systems more economically feasible. Because automation

TABLE 4.8
Uptime and Efficiency Comparison in Solar Panel Installation

System Type	Average Installation Time (h)	Defect Rate (%)	Daily Output (kWh)	Maintenance Need (per Year)	Cost per kW (US$)	Human Hours Required	System Downtime (%)
Manual only	7.3	3.9	7.6	5	1,120	28	6.2
Robotic assisted	4.2	1.1	9.4	2	840	11	2.3

reduces operating and maintenance expenses, renewable energy plants are more profitable. As a result, this draws in more and more capital to accelerate growth. The economy and sustainability of renewable energy projects have been completely transformed by automation. Automation increases operational efficiency, material savings, and labour efficiency, all of which reduce costs and increase the appeal of renewable energy projects to investors. It also accelerates technology adoption, which helps us transition to a low-carbon economy and quickly adjust to climate change. This combination of robots, AI, and the IoT will have a profound impact on the future of the renewable energy sector. Through increased economic feasibility and compatibility with clean energy solutions' sustainability goals, these technologies guarantee that renewable energy benefits both current and future generations.

THOUGHT BOX

WHO OWNS THE DATA IN SMART CONSTRUCTION SITES?

With massive data generated by IoT-enabled robots, who has the right to control, access, or monetize this information? Workers? Companies? Governments? How do we ensure privacy and equity in data governance?

4.6 Challenges and Future Directions

The construction sector is adopting the concepts of Construction 5.0, particularly the renewable energy projects; however despite all the benefits, this movement has several drawbacks. Together, they provide chances for sustainability and increased productivity, but they also bring with them practical difficulties, moral conundrums, and a growing need for staff upskilling. The future appearance of buildings as smarter and greener technology become more prevalent will also be covered in this chapter. There are moral and practical concerns when these new technologies are used in actual construction and renewable energy projects. The use of the recently created technology also raises several ethical concerns, such as the loss of jobs due to the automation of some tasks. The building of wind turbines and solar panels are two more examples. Although output has grown due to these technologies, workers who are unable to switch to alternative manufacturing lines are displaced. By 2030, for instance, more than 20% of construction-related employment may be affected by automation, according to the International Federation of Robotics. Nonetheless, it necessitates a thorough strategy in which the same personnel must be upskilled and re-skilled through early use of technology. The organizations are unable to sustain social inequity and mass unemployment through technical innovation and development of data security and privacy. By 5.0, there are also some significant problems with the employment of AI and IoT. This is since IoT devices gather and transmit vast volumes of data on the locations, equipment, and operational procedures of renewable energy projects, which they then privately store. If there are security breaches, there may be project delays, financial losses, and even safety risks. Such dangers would be decreased by robust cybersecurity solutions, including encryption, secure communication protocols, and frequent system audits.

Additional ethical concerns are brought up by data ownership and use. To safeguard the developers and staff, there must be explicit policies governing the protection of the interests of the individuals or parties involved in the project. From the procurement of raw materials to manufacturing, each of these stages involves energy-intensive procedures that emit greenhouse gases. With great power comes great responsibility. Robotics in construction must adhere to rigorous ethical, legal, and environmental standards. Figure 4.10 presents a governance model incorporating real-time oversight mechanisms and long-term regulatory frameworks. This ensures that Construction 5.0 technologies remain transparent, accountable, and aligned with sustainable development goals.

Most workers in today's market, however, lack the technical skills necessary for their positions. For example, more than half of construction workers

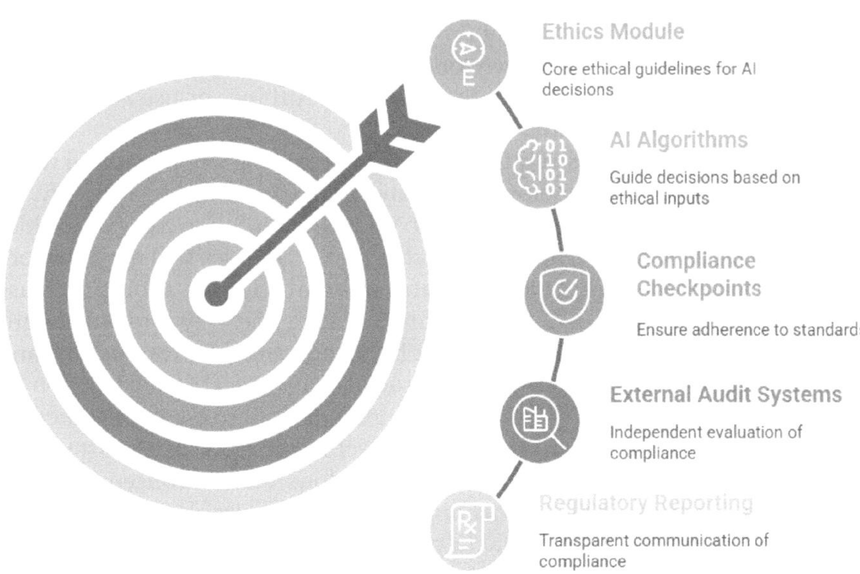

FIGURE 4.10
Ethical and legal integration in robotic construction.

would require retraining by 2030, according to a World Economic Forum assessment of the Construction 5.0 goals. Businesses and educational institutions may work together to make it happen. It takes comprehensive training in those areas, together with a culture of lifelong learning, to impart skills in data analysis, robots, AI, and the IoT. The online learning portals Coursera and edX, as well as the vocational training facilities, will make education extremely affordable. By enabling all segments of society to benefit from the opportunities provided by workers, exclusionary tendencies could be lessened. Women and members of other underrepresented groups should be prioritized to boost diversity in the Construction 5.0 workforce. In this sense, community involvement, mentorship programmes, and scholarships are ways that the government and all other sectors must work together to promote diversity.

The construction and education sectors must create a future workforce that collaborates well. Adding courses utilizing Construction 5.0 technology is one option to increase the number of courses offered at the university and technical institution. During these courses, academic knowledge will be applied through group projects and internships. For instance, Norway's NTNU has developed digital twinning and BIM programmes in response to the requirement for modern construction. A system like this links theoretical understanding to real-world applications. Through these activities, pupils would become more creative as they developed and prepared for future responsibilities. Besides upskilling via hands-on instruction, employees require on-the-job training in addition to formal education to adjust to technological advancements. Training programmes that allow workers to pick up new skills while working on projects might be created by construction companies. For example, the worker will be given hands-on training on how to handle robotic equipment to build wind turbines or use AI-powered project management software.

These initiatives also assist businesses in maintaining key personnel while continuing to operate. Employee relocation also goes more smoothly. In order to overcome the obstacles on the path to Construction 5.0, stakeholder assistance is crucial. Government, industry, and academics will need to work together as best practices, guidelines, and standards for the appropriate use of technology in construction are defined. Along the way, breakthroughs will be made in the areas of autonomous buildings, smart infrastructure, and renewable energy integration. Another crucial component of Construction 5.0 design will be sustainability. Future developments will consequently include more ecologically friendly materials, energy-efficient robots, and sustainable production techniques. Research should be done on the consequences of technology adoption on the environment, which the circular economy concept should make less detrimental or even beneficial. The construction sector needs to adopt a new culture of lifelong learning to keep up with the rapid changes that are occurring. Governments and businesses may offer learning environments where skills will be created to assist workers in adjusting to the rapid changes. These contexts might include online learning environments, accreditation programmes, and knowledge exchange between individuals.

This implies that ethical practices should guide the creation and application of Construction 5.0 technology. Therefore, there should be more openness, inclusivity, and accountability from developers to prevent technological innovation from going against the prevailing trend of society acceptance or from enriching a select few. There are both possible benefits and obstacles as the construction sector embraces Construction 5.0, particularly in relation to renewable energy. If the construction sector addressed ethical concerns, developed collaborative skills, and filled in skill gaps, these challenges would be resolved. By combining technology innovation with upskilled personnel, Construction 5.0 has the potential to improve the sector and make it stronger, more efficient, and sustainable. In such case, the building industry's proactive strategy will be standard for a more intelligent and ecologically conscious world.

4.7 Conclusion

Robotics has unquestionably transformative potential to advance green energy objectives. This will assist global environmental initiatives and alter the dynamics of the construction industry. Robots, the foundation of Construction 5.0, are revolutionizing renewable energy projects through accuracy, efficacy, and safety when paired with AI and IoT. Technology integration is not only speeding up construction but also enabling the deployment of renewable energy infrastructure at a significantly faster rate than previously believed. Essentially, these two issues of sustainability and efficiency can be resolved by robots. Labour-intensive jobs like installing solar panels or building wind turbines could be automated by robots. It significantly reduces waste, material costs, and construction schedules. It allows the globe to advance in renewable energy sources and makes renewable energy projects more accessible worldwide. Remote areas that were previously inaccessible, like offshore wind farms or solar farms, are also handled by robotics. Because of their remoteness, these areas encourage the use of renewable energy sources.

Cooperation between humans and robots is the other relevant component. Collaboration between humans and robots is highly valued in these projects because Construction 5.0 incorporates human decision-making abilities into a robot's precision and resilience. Since operations in renewable energy projects are typically complicated and require mechanical precision, creativity, and flexibility, this synergy is being heavily utilized. Telepresence robots, for instance, enable experts to keep an eye on operations from a safe distance, guaranteeing security and safety in risky circumstances. Exoskeletons also allow workers to be more productive since they put less physical strain on their bodies. By achieving considerable cost reductions through integration with contemporary technologies, renewable energy may now compete with traditional fossil fuels. In renewable energy projects, robotics enhances the financial and ecological advantages. It is lowering the carbon footprint associated with building. From that perspective, robotics helps to achieve sustainability goals. Deploying autonomous, energy-efficient machinery and IoT-enabled predictive maintenance systems is necessary to meet international climate targets in renewable energy projects. These solutions use even less resources and emit fewer emissions. However, robots present several difficulties for applications in the construction industry. Data privacy and employment displacement are ethical issues for responsible and equitable deployment. Construction 5.0 will rely heavily on its staff, which requires expertise, adaptability, and the ability to use cutting-edge technologies. To meet the expectations of a technologically sophisticated industry, the workforce needs to be prepared, which calls for investments in lifelong learning, upskilling, and training. The continuous advancement of robotics in renewable energy projects will pave the way for future innovations and sustainability. Construction is well-positioned to take the lead in accomplishing renewable energy targets, thanks to advancements in AI, IoT, and collaborative robotics. Construction 5.0 has the potential to drastically alter the process of building a more environment-friendly and sustainable future by utilizing these technologies and resolving pertinent issues. The promise and transformative potential of robotics are the foundation of contemporary green energy projects.

Chapter Review: Chapter 4 – Robotics in Construction 5.0

Multiple Choice Questions

Test your understanding of the key concepts from this chapter. Each question is followed by four options, only one is correct.

1. What is a primary focus of Construction 5.0 compared to earlier industrial revolutions?
   A. Speed of production
   B. Human-centric automation
   C. Heavy manual labour
   D. Sole reliance on AI

2. Which of the following represents a wearable robotic system used on construction sites?
   A. Telepresence drone
   B. AI-powered concrete mixer
   C. Exoskeleton suit
   D. Robotic crane
3. What is the main advantage of telepresence robots in hazardous construction zones?
   A. Faster physical labour
   B. Real-time collaboration without physical risk
   C. Physical strength enhancement
   D. Increased fossil fuel use
4. The integration of AI and IoT on smart construction sites enables:
   A. Elimination of human workers
   B. Better weather forecasting
   C. Real-time decision-making and process automation
   D. Traditional project management systems
5. What makes context-aware robotic systems unique in construction environments?
   A. Repetitive programming
   B. Manual override dependency
   C. Ability to adapt to dynamic environments
   D. Limited sensor input
6. Which of the following is a direct sustainability benefit of robotics in construction?
   A. Increased human labour hours
   B. Lower battery usage
   C. Precision in material usage and reduced waste
   D. Less reliance on digital technology
7. The main function of exoskeletons in the construction sector is to:
   A. Replace engineers
   B. Enhance physical performance and reduce fatigue
   C. Perform land surveys
   D. Increase manual labour quotas

8. How do robots contribute to wind turbine assembly?
   A. Managing legal compliance
   B. Financial modelling
   C. Precision lifting and component alignment
   D. Surface painting
9. Manual solar panel installations are often affected by:
   A. AI failures
   B. Low material cost
   C. Fatigue and alignment errors
   D. Excessive use of robotics
10. A skill transition pathway helps workers to:
    A. Retire earlier
    B. Avoid upskilling
    C. Adapt to new robotic roles with training
    D. Increase on-site risks
11. Ethical oversight in robotics is essential to:
    A. Accelerate project deadlines
    B. Lower insurance premiums
    C. Ensure safety, equity, and regulatory compliance
    D. Increase wages automatically
12. A co-bot on a construction site is best described as a robot that:
    A. Works autonomously with no oversight
    B. Manages HR processes
    C. Shares tasks collaboratively with humans
    D. Removes human intervention
13. What is a key advantage of using digital twins in Construction 5.0?
    A. Power storage
    B. Blueprint encryption
    C. Simulating and optimizing real-time processes
    D. Replacing building permits
14. IoT sensors in robotic construction sites primarily help to:
    A. Predict workforce stress
    B. Advertise products
    C. Monitor structural and environmental conditions
    D. Encrypt communications

15. The long-term success of robotic construction depends on:
    A. Avoiding automation
    B. Centralizing labour
    C. Human–robot synergy and ethical deployment
    D. Decreasing human education

Answer Key

Question No.	Correct Option	Question No.	Correct Option
1	B	9	C
2	C	10	C
3	B	11	C
4	C	12	C
5	C	13	C
6	C	14	C
7	B	15	C
8	C		

References

Adekanbi, M. L. (2021). Optimization and digitization of wind farms using Internet of Things: A review. *International Journal of Energy Research*, *45*(11), 15832–15838.

Anitha, D., Kolangiammal, S., Lalitha, K., Valli, K. V. N., Ganesh, E., & Muthulekshmi, M. (2024). IoT-enabled exoskeletons for firefighters using reinforcement learning for adaptive support in emergency situations. 2024 Second International Conference on Intelligent Cyber Physical Systems and Internet of Things (ICoICI),

Barata, J., & Kayser, I. (2023). Industry 5.0: Past, present, and near future. *Procedia Computer Science*, *219*, 778–788.

Boschetti, G., Faccio, M., & Granata, I. (2022). Human-centered design for productivity and safety in collaborative robots cells: A new methodological approach. *Electronics*, *12*(1), 167.

Çelik, D., Meral, M. E., & Waseem, M. (2022). Investigation and analysis of effective approaches, opportunities, bottlenecks and future potential capabilities for digitalization of energy systems and sustainable development goals. *Electric Power Systems Research*, *211*, 108251.

Dindorf, R., & Wos, P. (2024). Challenges of robotic technology in sustainable construction practice. *Sustainability*, *16*(13), 5500.

Duncan, G. M., Mugabe, K., Zimmerman, K. A., Sibaja, V., & Luis, M. (2018). Adapting a Culture for Performance Management at the Nevada Department of Transportation. NDOT Research Report. Report No. 365-16-803. Nevada Department of Transportation.

Firoozi, A. A., Hejazi, F., & Firoozi, A. A. (2024). Advancing wind energy efficiency: A systematic review of aerodynamic optimization in wind turbine blade design. *Energies, 17*(12), 2919.

Gajdzik, B., Nagaj, R., Wolniak, R., Bałaga, D., Žuromskaitė, B., & Grebski, W. W. (2024). Renewable energy share in European industry: Analysis and extrapolation of trends in EU countries. *Energies, 17*(11), 2476.

Hassan, A. A., El Habrouk, M., & Deghedie, S. (2020). Renewable energy for robots and robots for renewable energy: A review. *Robotica, 38*(9), 1576–1604.

Huang, S., Wang, B., Li, X., Zheng, P., Mourtzis, D., & Wang, L. (2022). Industry 5.0 and society 5.0: Comparison, complementation and co-evolution. *Journal of Manufacturing Systems, 64*, 424–428.

Knips, C., Bertling, J., Blömer, J., & Janssen, W. (2014). FabLabs, 3D-printing and degrowth: Democratisation and deceleration of production or a new consumptive boom producing more waste? Fourth International Conference on Degrowth for Ecological Sustainability and Social Equity.

Kolade, O., & Owoseni, A. (2022). Employment 5.0: The work of the future and the future of work. *Technology in Society, 71*, 102086.

Lai, X., Wang, S., Guo, Z., Zhang, C., Sun, W., & Song, X. (2021). Designing a shape – performance integrated digital twin based on multiple models and dynamic data: A boom crane example. *Journal of Mechanical Design, 143*(7), 071703.

Linner, T., & Bock, T. (2012). Evolution of large-scale industrialisation and service innovation in Japanese prefabrication industry. *Construction Innovation, 12*(2), 156–178.

Liu, S. B., & Althoff, M. (2020). Optimizing performance in automation through modular robots. 2020 IEEE International Conference on Robotics and Automation (ICRA).

Mehta, P. (n.d.) Accelerating Solar Revolution: Advances in Photovoltaic Cells and IoT Energy Integration.

Mitchell, D., Blanche, J., Harper, S., Lim, T., Gupta, R., Zaki, O., Tang, W., Robu, V., Watson, S., & Flynn, D. (2022). A review: Challenges and opportunities for artificial intelligence and robotics in the offshore wind sector. *Energy and AI, 8*, 100146.

Musarat, M. A., Irfan, M., Alaloul, W. S., Maqsoom, A., & Ghufran, M. (2023). A review on the way forward in construction through Industrial Revolution 5.0. *Sustainability, 15*(18), 13862.

Ness, S., Shepherd, N. J., & Xuan, T. R. (2023). Synergy between AI and robotics: A comprehensive integration. *Asian Journal of Research in Computer Science, 16*(4), 80–94.

PV Magazine USA. (2023). Utility-scale solar installation goes automated. pv-magazine.com

Rane, N., Choudhary, S., & Rane, J. (2023). Leading-edge wearable technologies in enhancing personalized safety on construction sites: A review. *SSRN Electronic Journal*. DOI: 10.2139/ssrn.4641480.

Renewable Energy World. (2025). AES sends world's first AI-powered solar robot fleet to California project. Factor This.

Sabet, M., Orand, M., & McDonald, W. D. (2021). Designing telepresence drones to support synchronous, mid-air remote collaboration: An exploratory study. Proceedings of the 2021 CHI Conference on Human Factors in Computing Systems.

Sandler, B. Z. (1999). *Robotics: Designing the mechanisms for automated machinery*. Elsevier.

Sepasgozar, S. M., Costin, A. M., Karimi, R., Shirowzhan, S., Abbasian, E., & Li, J. (2022). BIM and digital tools for state-of-the-art construction cost management. *Buildings, 12*(4), 396.

Simas, M., & Pacca, S. (2014). Assessing employment in renewable energy technologies: A case study for wind power in Brazil. *Renewable and Sustainable Energy Reviews, 31*, 83–90.

Stephen, S. S., Oke, A. E., & Aigbavboa, C. O. (2024). Integration of Industry 5.0 principles in stealth construction: Leveraging emerging technologies for efficiency and sustainability. Industry 5.0 and emerging technologies: Transformation through technology and innovations (pp. 181–197). Springer.

Stipancic, T., Jerbic, B., & Curkovic, P. (2016). A context-aware approach in realization of socially intelligent industrial robots. *Robotics and Computer-Integrated Manufacturing, 37*, 79–89.

Su, Y.-P., Chen, X.-Q., Zhou, C., Pearson, L. H., Pretty, C. G., & Chase, J. G. (2023). Integrating virtual, mixed, and augmented reality into remote robotic applications: A brief review of extended reality-enhanced robotic systems for intuitive telemanipulation and telemanufacturing tasks in hazardous conditions. *Applied Sciences, 13*(22), 12129.

Ukoba, K., Olatunji, K. O., Adeoye, E., Jen, T.-C., & Madyira, D. M. (2024). Optimizing renewable energy systems through artificial intelligence: Review and future prospects. *Energy & Environment, 35*(7), 3833–3879.

Walker, M., Chen, Z., Whitlock, M., Blair, D., Szafir, D. A., Heckman, C., & Szafir, D. (2021). A mixed reality supervision and telepresence interface for outdoor field robotics. 2021 IEEE/RSJ International Conference on Intelligent Robots and Systems (IROS).

Xu, X., Lu, Y., Vogel-Heuser, B., & Wang, L. (2021). Industry 4.0 and Industry 5.0: Inception, conception and perception. *Journal of Manufacturing Systems, 61*, 530–535.

Yavari, F., & Pilevari, N. (2020). Industry revolutions development from Industry 1.0 to Industry 5.0 in manufacturing. *Journal of Industrial Strategic Management, 5*(2), 44–63.

Yitmen, I., Almusaed, A., & Alizadehsalehi, S. (2024). Facilitating Construction 5.0 for smart, sustainable and resilient buildings: Opportunities and challenges for implementation. *Smart and Sustainable Built Environment*. DOI: 10.1108/SASBE-04-2024-0127.

5

Building Information Modelling and the Fifth Industrial Revolution: Collaborative Construction of Renewable Energy Facilities

5.1 Building Information Modelling

Building information modelling (BIM) is a visual representation of project's physical and functional characteristics. It is a knowledge base domain that promotes informed decision-making for a project throughout its life cycle. According to US National Building Information Model Standard Project Committee, 2016, BIM provides detailed knowledge from the design to demolition of a project. It is more than a simple model but a process that involves creating, managing, and sharing digital 3D visuals with comprehensive details in it. The major advantage of this visualization is that all the stakeholders can access and utilize this enriched representation to achieve all the milestones from design to execution.

This represents the latest innovation in the construction sector, integrating the collaborative design, building, and operation of a project via a cohesive system of 3D models rather than separate design drawings. BIM integrates personnel and technology to optimize time and cost, enhancing efficiency in the construction of skyscrapers, hospitals, office complexes, and residential structures. BIM is not merely a collection of software or a basic 3D model. It encompasses both the model elements and the extensive information constituting the project, together with the procedure for disseminating that information to other stakeholders. In contrast to earlier workflows that depended on several file formats and disjointed processes, which rapidly fell out of sync with modifications, BIM workflows provide a more dynamic and synchronized method of project management. The formation of BIM can be found in Figure 5.1, which shows that initially they started with a computer-aided design (CAD) system, then a graphical user interface (GUI), 2D CAD drawing, and now all those previous technologies have been merged and upgraded as BIM.

Originally known as the building description system (BDS), the idea behind BIM emerged in the 1970s. The first use of the term 'building model' appeared in an article on CAD and other forms of CAD in 1985. An article

DOI: 10.1201/9781003599272-5

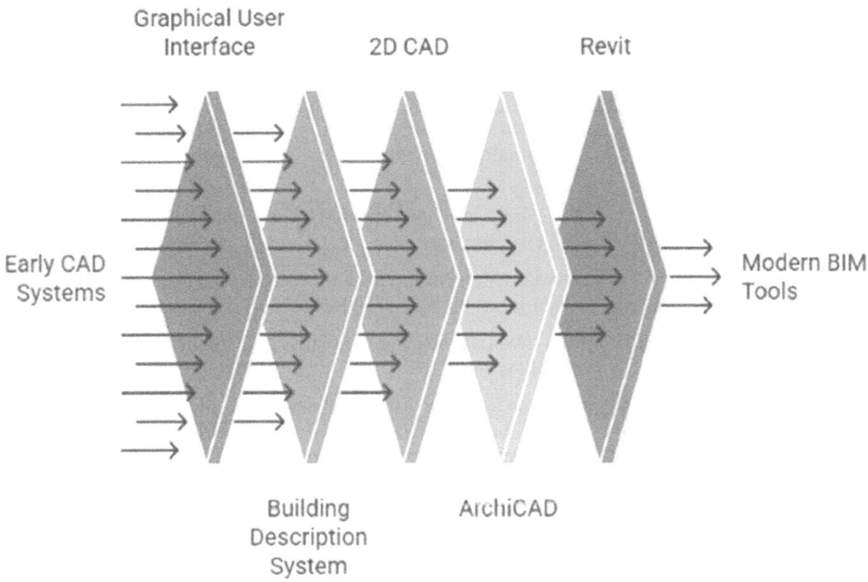

FIGURE 5.1
BIM evolution.

discussing construction automation first used the phrase 'building information model' in 1992. Ten years later, the acronym BIM and the phrases building information modelling and building information model were first used, and gained widespread usage. In 2002, Autodesk released a paper titled 'BIM', leading to the involvement of other software developers and vendors in the sector, thereby standardizing the term to denote the digital representation of the building process. Due to increased demand and unsustainable energy supply, energy efficiency has become increasingly vital. Energy markets inherently possess significant ramifications, affecting everything, from home finances to international diplomacy.

5.2 Historical Evolution of BIM

5.2.1 From 2D CAD to 3D Collaboration

BIM emerged in the 1970s as a basic digital design instrument and progressed into a collaborative platform uniting stakeholder throughout a project's life cycle. Table 5.1 shows the evolution of BIM along the timeline, technologies that have been used, innovation, and their impact.

Buildings account for around 40% of worldwide energy use (International Energy Agency, 2019). Buildings, owing to their substantial energy usage,

TABLE 5.1
BIM Technologies Evolution

Decade	Technology	Key Innovation	Impact
1970s	Building description system	Initial computer depiction of building components	Laid groundwork for parametric modelling
1990s	3D modelling software	Tools like ArchiCAD enabled basic 3D visualization	Shifted focus from 2D blueprints to 3D
2000s	BIM standardization	Standards for interoperability were introduced by Autodesk Revit	Enabled multidisciplinary collaboration
2010s	Cloud-based BIM	Real-time data sharing via platforms like BIM 360	Enhanced global project coordination
2020s	AI-integrated BIM	Machine learning (ML) for clash detection and energy modelling	Automated design optimization

are at the forefront of energy efficiency research. Consequently, the proper execution of building energy efficiency policies will significantly benefit both economically and environmentally. It is a digital approach that streamlines the planning, design, construction, and management of built environments via a cohesive and data-intensive platform. BIM, created for visualizing architectural designs, has transformed into a fundamental element of integrated project delivery, facilitating cooperation among stakeholders and supporting the entire life cycle of infrastructure assets. As time passes, various milestones have been shifted through BIM. Table 5.2 shows the milestones along with the time and their significance.

In present sustainable architecture and engineering, there is an increasing focus on integrating renewable energy systems into structures. The difficulty of this integration, however, resides in the intricacy of the design and implementation process. The industry is increasingly adopting BIM to address this difficulty, a technical advancement that has transformed the design, construction, and management of building infrastructure. As a leading entity in BIM technology, we have dedicated over a decade to advancing boundaries and establishing new benchmarks in the BIM domain.

5.3 Digital Representation

BIM elements significantly influence the endeavour to digitally depict the complete life cycle of a constructed entity. With this accurate construction model, the project team can remain aligned and convey important updates when needed in a clear and timely way. Moreover, a meticulously designed BIM model enables construction managers to evaluate several alternative

TABLE 5.2

Timeline of BIM

Year	Milestone	Significance
1975	Building description system (BDS)	Regarded as the inaugural BIM platform, established a fundamental pattern for BIM, albeit restricted to design components
1977	GLIDE	GLIDE advanced BDS by providing superior visualization and enabling estimating from designs, although it remained confined to the preconstruction phase
1989	Building product model (BPM)	Launched as the inaugural BIM product, facilitating the integration of design, estimation, and project delivery phases by converting project information into interconnections across construction activities
1995	Generic building model (GBM)	Developed to enhance the integration of information with construction operations over a project's life cycle
2000	Building information modelling	The acronym BIM was introduced and it gained prominence as a transformative concept resulting from technical advancements, which prompted numerous BIM mandates and heightened utilization of 3D modelling tools across the AEC sectors
2013	BuildingSMART	BuildingSMART International introduced the Industry Foundation Classes (IFC) standard to facilitate data exchanges among various software systems

scenarios to see the complete planned sequence of the project. This visual depiction can be shared with the client and other stakeholders, serving as a foundation for the following actions. It is essential to emphasize that data must precede 3D. Visual depiction is only as effective as the data utilized in its production. Consequently, the distinction between 2D and 3D designs and blueprints is not the primary concern at this juncture. The construction sector must prioritize digital adoption before it can fully explore the prospects presented by 3D representations of constructed structures. The worldwide construction industry is leading a technological and cultural shift to meet the urgent challenges of climate change, digital innovation, and sustainable infrastructure development. One of the most revolutionary tools facilitating this transition is BIM, a process-oriented methodology that uses digital representations of the physical and functional attributes of constructed assets. BIM transcends conventional 2D and 3D modelling, encompassing the entire life cycle of a project from conceptual design and construction to operations, maintenance, and final decommissioning. This cohesive strategy promotes collaboration among architects, engineers, contractors, and facility managers, facilitating enhanced decision-making, less project risk, and increased overall efficiency.

> *The evolution of BIM reflects the construction industry's transition from fragmented processes to integrated digital ecosystems.*

> *'From 2D CAD to AI-driven BIM: How have historical technological constraints influenced contemporary collaborative workflows?' What insights may future innovators derive from the repeated evolution of BIM?*

By integrating BIM into the project, planners can identify optimal spots. Architects may provide more precise designs with reduced errors, minimized waste, and enhanced alignment with the owner's vision. Engineers can enhance collaboration with architects and other engineering fields, hence augmenting the trustworthiness of their designs. Contractors can ensure that constructability issues are identified early when modifications are less costly to implement. Ultimately, proprietors will be able to utilize the models for an extensive facility and asset management programme well into the future. Embracing the benefits of BIM necessitates investment in personnel, procedure, and technology. Embracing these changes with an optimistic perspective will expedite the shift and enable organizations to swiftly attain the productivity and quality enhancements achievable through BIM.

5.4 Industry 5.0

The rise of the Fifth Industrial Revolution (Industry 5.0) signifies a new phase in industrial progress, focusing on human–machine collaboration, sustainability, resilience, and intelligent decision-making. In contrast to Industry 4.0, which emphasizes automation and digitalization, Industry 5.0 adopts a human-centric paradigm that incorporates technology such as artificial intelligence (AI), digital twins (DT), Internet of Things (IoT), and robotics to augment, not supplant human abilities.

The Fifth Industrial Revolution (Industry 5.0) has emerged as the subsequent phase in the evolution of industrial practice, after the automation-driven trajectory of Industry 4.0. Industry 5.0 introduces human-centric paradigms, while Industry 4.0 focused on cyber-physical systems like IoT, and data analytics. It focused on the combination of human creativity and intelligence of a machine like, AI, DT, robotics, and augmented reality (AR). The main objective is not to eliminate human but to combine their capability with improved working environment. Industry 5.0 prioritizes sustainability, resilience, personalization, and the ethical integration of technology. The reasons why numerous aspects of the Industry 4.0 idea are perceived by many as insufficient, unfinished, and in need of enhancement should be readily apparent. In one of his tweets, Elon Musk stated: 'Yes, excessive automation at Tesla was a mistake. To be precise, my mistake. Humans are underrated'. Also, Frost & Sullivan, a reputable American business consulting firm, defined its conversion to Industry 5.0 by stating: 'Industry 4.0 relies heavily on automation and has been intimidating workers on factory shop

TABLE 5.3

Different Aspects of Industry 4.0 and 5.0

Aspect	Industry 4.0	Industry 5.0
Core focus	Automation, IoT, big data	Human-centric design, sustainability, resilience
Workforce role	Humans are being replaced by robots	Humans augmented by cobots and AI
Sustainability	Limited emphasis on eco-efficiency	Net-positive environmental impact, circular economy
Key technologies	Cyber-physical systems, predictive analytics	DT, collaborative robotics, blockchain

floors. Frost & Sullivan envisions a futuristic scenario of the next big thing, Industry 5.0, which will bring back empowered humans to the shop floor'.

Industry 5.0 emphasizes human–machine collaboration, sustainability, and resilience, contrasting with Industry 4.0's focus on automation. Table 5.3 differentiates between Industry 4.0 and Industry 5.0, based on four different aspects.

5.5 Industry 5.0 Strategies

Industry 5.0 is underpinned by three strategies – human-centric, resilience, and sustainability – which can also be seen in Figure 5.2.

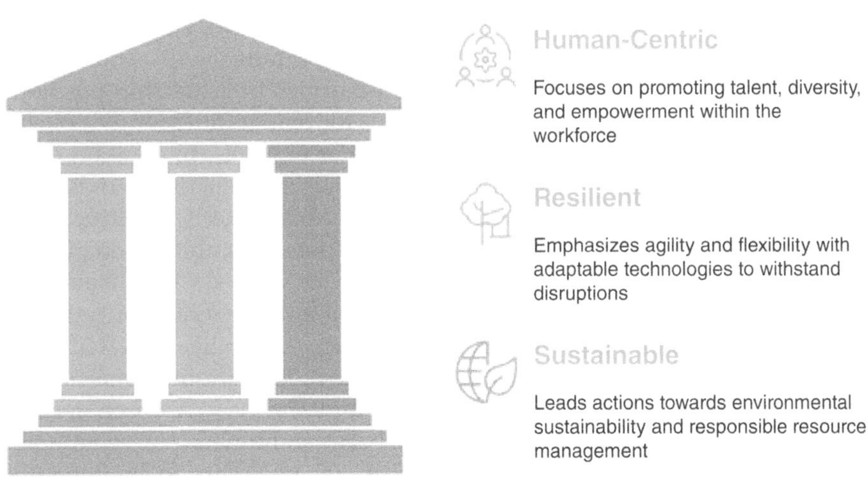

FIGURE 5.2
Strategies of Industry 5.0.

5.5.1 Human-Centric Strategy

Industry 5.0 encompasses a strategy that repositions individuals from mere resources to authentic assets. This signifies that organizations will prioritize serving individuals rather than individuals serving organizations. Rather than solely utilizing talent to establish a competitive edge and deliver value to consumers, Industry 5.0 shifts its focus to also generate additional value for employees, thereby attracting and retaining top talent. Human-centric approach needs to be a balanced approach, because in its absence, machine or technology cannot reach up to its full potential. With the implementation of this strategy, society can examine critical socio-environmental data and AI-based decision-making in the favour of global sustainable development. Involving human-centric approaches into industrial systems will overcome current challenges and improve industry efficiency and productivity. It requires intelligent robots that can understand this complex relationship with human and machines in complex environments. Looking at the big picture, it becomes clear that trust-based connections are crucial for a human-centric society. They enable people to engage efficiently and autonomously maintain the security and stability of the social infrastructure, that is, communication networks.

The main aim of Industry 5.0 is to involve this human-centric approach in those businesses or industries which did not perform well in era of Industry 4.0. The collaboration between humans and robots is crucial for generating additional employment, enhancing production and efficiency, and recruiting and retaining talent. For technology-centric enterprises, human capital represents the most critical asset, necessitating that businesses adapt to the evolving requirements of the increasing cohort of Gen-Z and Millennials. Younger generations of workers prioritize environmental and social considerations in their employer selection, which may encompass support for community initiatives, flexible working arrangements, and the promotion of historically underrepresented minorities in leadership positions. Worldview on human-centric approach calls for reimagining for the collaboration between the machines and workers. For example, integration of DT and cobots can enhance the performance of static manufacturing industries.

When it comes to scientific advancements in building projects, Construction 5.0 is all about human-centric technology, which puts an emphasis on people's experiences and uses dynamic approaches. This marks a shift from solutions driven by technology to technology that enhances human capacities and welfare. This includes developments such as intelligent systems that enhance human skills rather than replace them, human–machine collaboration, and AR applications for workers. Simultaneously, research encounters challenge in maintaining a harmony between human intervention and automated technologies. To accomplish this, it is necessary to combine BIM with non-BIM components, making use of the features of both IT and non-IT systems. Additionally, it is necessary to link the real and virtual worlds to guarantee smooth integration all the way through the design, construction, and delivery stages. It is imperative that the AEC industry's future research endeavours investigate multiple interrelated subjects. Themes include integrated design with BIM and DT analysis; cloud computing for collaborative design; modular integrated construction (MiC) with 4D printing and collaborative robots (cobots); MiC with mass customization; smart contracts and blockchain applied; and big data analytics deployed in the supply chain.

Human-centric technology in Construction 5.0 emphasizes aligning technological innovation with human needs, experiences, and capabilities throughout construction processes. This marks a shift from purely technology-driven solutions to those that enhance and support human performance and well-being. Key innovations include human–machine collaboration, AR tools for workers, and intelligent systems designed to complement – not replace – human roles. A major research challenge lies in achieving a harmonious integration of physical labour and automation, requiring the combination of both BIM and non-BIM elements, and the effective use of IT and non-IT platforms. This integration must bridge the physical and digital environments to ensure seamless coordination across design, construction, and delivery phases. Moving forward, research in the AEC sector must explore various integration themes such as the role of BIM and DT in unified design processes, the use of cloud computing for collaborative development integrated construction (MiC) using cobots and 4D printing, large-scale customization within MiC, smart contracts and blockchain deployed, and big data analytics used to construction supply chain management.

5.5.2 Resilience

Global issues, like as the COVID-19 pandemic and international supply shortages, have far-reaching consequences due to the world's growing interconnection. While most enterprises focus on enhancing efficiencies and maximizing earnings, these elements do not augment resilience. Indeed, there exists a view that an emphasis on agility and flexibility may render firms less durable rather than more. Instead of concentrating on expansion, profit, and efficiency, more resilient organizations would prioritize anticipating and responding to crises to maintain stability during hard periods. Resiliency denotes a system's capacity to maintain or swiftly return to a stable condition during and after a considerable disruption associated with geopolitical events and natural disasters, including the COVID-19 pandemic, or in the context of persistent significant pressures.

The capacity for resiliency is an acknowledged fundamental competency of Industry 5.0. Industry 5.0 emphasizes not only the capacity of enterprises to navigate external uncertainties, including market fluctuations, supply chain disruptions, and customer variability, but also the resilience of broader industrial systems, encompassing the entire industrial chain and even the industrial framework of a nation or region against unforeseen risks.

Within the framework of In Construction 5.0, resilience refers to the industry's ability to adapt and recover from challenges or disruptions. This may entail integrating technologies that facilitate rapid adaptations to evolving circumstances, resilient techniques, and systems designed to predict and manage hazards. It also includes the ability to recover from unforeseen setbacks and sustain ongoing operations. In recent years, the AEC business has demonstrated notable resiliency. Organizations that adopt and adjust to emerging technologies, particularly during the epidemic, are positioning themselves to surpass their rivals.

5.5.3 Sustainability

Industry 5.0 expands the concept of sustainability beyond merely reducing, minimizing, or mitigating climate damage to actively strive for positive transformation. Occasionally termed 'Net Positive', this objective seeks to enhance global conditions by encouraging firms to contribute to solutions rather than perpetuating issues or merely engaging in superficial commitments to sustainability through 'greenwashing'. All previous industrial revolutions demonstrate a collective acknowledgement of the superior advantages of sustainability. Sustainable manufacturing is essential for the sustainable advancement of society in areas including sources of renewable energy (Scharmer et al., 2023). Sustainable manufacturing serves as a framework within the United Nations' Sustainable Development Goals, namely in the areas of 'Responsible Consumption and Production' and 'Industry, Innovation, and Infrastructure'. Sustainability relies on the innovation of products and services.

To achieve mass customization in products and services, sustainable manufacturing is needed, which is based on not centralized but decentralized integration of socially connected resources. Sustainability majorly depends on three key pillars: economic, environment, and society. In contrast to previous industrial revolutions, Industry 5.0 emphasizes human-centric societal approaches, leading to a gap among the three pillars of sustainability, which should ideally be harmonized. The economic dimension of sustainability is crucial in the technological reorganization necessary to achieve the Industry 5.0 goal. The sustainable implementation of Industry 5.0 involves timely adjustments of one aspect within the three pillars at various phases to enhance the global development of Industry 5.0, resulting in increased quantity, accelerated speed, improved quality, and cost efficiency.

The motive of Industry 5.0 is to promote an integrated approach in construction, while Construction 5.0 is focused on sustainability by minimizing the environmental impacts and long-term resilience, which requires the use of eco-friendly materials, efficient design, minimum waste production strategies, and the involvement of green technology. The development of sustainable industrial products is the prime focus of Industry 5.0, by using renewable energy sources and waste reduction. Also, it is important to have real-time monitoring and decision-making, which can be possible with the use of a variety of sensors so that it can increase efficiency with efficient utilization of resources. The Industrial Revolution 5.0 fundamentally involves the establishment of a more inclusive industrial framework that ensures accessibility for all individuals. This entails the utilization of advanced technologies, including virtual and AR, to deliver training and support to employees, alongside the creation of adaptive production processes that cater to workers with varying abilities.

> *'Can human-centricity, resilience, and sustainability coexist without trade-offs? Does improving worker safety using cobots undermine cost efficiency? What metrics are utilized to assess 'success' within a human-centric industrial ecosystem?*
>
> *If Industry 5.0 emphasizes human–machine collaboration, how can industries guarantee that technological progress does not result in job displacement? Can ethical frameworks reconcile innovation with employee welfare?*
>
> *'Given that buildings constitute 40% of global energy consumption, is it feasible for BIM to independently attain net-zero objectives?' What systemic alterations are required in addition to technology solutions?*

5.6 BIM and Industry 5.0

The convergence of BIM with Industry 5.0 technologies offers a significant opportunity to revolutionize the conception, construction, and operation of infrastructure, particularly renewable energy plants. With the increasing global need for renewable energy, there is a corresponding necessity for efficient, sustainable, and collaborative construction methodologies. Renewable energy initiatives, including wind farms, solar installations, hydropower plants, and biomass facilities, are intrinsically intricate and multidisciplinary. The integration of BIM with human-centric technology facilitates resource optimization, life cycle management, predictive maintenance, and real-time collaboration among various teams and locations.

For many organizations, Industry 4.0 remains the forthcoming priority or the trend to which they are presently aligning their strategies. This represents the fourth industrial revolution, driven by developments in information technology. Key components include automation, robotization, big data analytics, intelligent systems, virtualization, AI, machine learning (ML), and the IoT. Industry 5.0 is a paradigm for redefining the future of energy, manufacturing, transportation, and supply chains, enhancing and expanding upon the significant foundation established by the vision of Industry 4.0. To bring the human element to the concept of digital transformation, Industry 5.0 uses cobots and AI. Economic, environmental, and social repercussions all are part of the 'triple bottom line' that the European Commission and other governments are pushing for with Industry 5.0, integrating an environmental, social, and governance (ESG) perspective to counterbalance traditionally technology-driven and economically motivated decisions. The architecture, engineering, and construction (AEC) sector is transforming, mentioning the need to integrate Industry 5.0 into the current construction strategies. As Industry 5.0 emphasized on human-centric, sustainable and resilient approach, this integration has changed the experience of construction industry. This integration is also important to promote intelligent, sustainable, and resilient construction.

5.7 Industry 5.0 versus Industry 4.0

A revolution in industry which focuses on involving advanced technologies and innovation within the domain of manufacturing is known as Industry 4.0. It has shown noticeable advantages in automating works through the help of machines but at the same time it has reduced human capabilities. On the other hand, Industry 5.0 worked towards rectifying this disparity by combining computing technologies like AI to guarantee a human presence so that the capabilities of human can also be increased with the increment in automation of processes. This shift from 4.0 to 5.0 has improved the design and administration processes, enhanced sustainability, and involves intelligent systems that work with human collaboration. This transition has also come with benefits and new challenges, asking companies and humans to adjust themselves according to the requirement of this system. Industry 5.0 seeks to combine human intelligence and creativity with computing capabilities. As compared to Industry 4.0, this emerging framework focused on collaboration and socio-economic transformation of the industry, where machines and human work together.

5.7.1 Sustainability

Industry 5.0 aims to establish a manufacturing ecosystem that is sustainable, resilient, and centred on human needs. A growing number of investors and regulators are placing significant emphasis on reducing carbon emissions and their associated environmental impacts; firms must assess their resource footprint that underpins their industrial processes. This might involve looking into where the resources come from, how much waste is made, how bad it is for the environment, how efficient the processes are, and where the energy comes from. Numerous organizations have pledged to reduce fossil fuel usage and the creation of clean energy sources. The utilization of novel materials and composites, particularly those not derived from petroleum, can mitigate environmental effects while enhancing smart manufacturing practices, including recycling and repurposing, to fulfil the goals of socio-ecological sustainability. The concept of Industry 5.0 extends beyond the limitations of 'industry'. It pertains to all sectors and organizations imaginable. This indicates that its applicability is considerably broader than that of Industry 4.0. Consequently, when examining the ramifications of Industry 5.0 for strategy, it is essential to have a comprehensive and universal viewpoint applicable to all sectors.

The notion of Industry 5.0 is rather novel. The European Union Industry 5.0 envisions an industrial paradigm that transcends mere efficiency and productivity, emphasizing the industry's function and contribution to society. It prioritizes the worker's wellness inside the production process and uses innovative technology to foster success that transcends mere employment and economic growth, all while honouring the planet's production constraints. It enhances the Industry 4.0 paradigm by explicitly positioning research and innovation to facilitate the transition towards a sustainable, human-centric, and resilient European industry. 'Industry 5.0 recognizes the power of industry to achieve societal goals beyond jobs and growth to become a provider of prosperity, by making production respect the boundaries of our planet and placing the well-being of the industry worker at the centre of the production process' (European Commission in the R&I Policy Brief, 2021).

Numerous countries are allocating resources towards research and development to advance environmentally sustainable construction methods. Consequently, upgrading the construction industry will ensure sustainable advancements. Value chain has been transforming with the implementation of some technologies like AI, 3D printing, augmented and virtual reality (VR), the IoT, and blockchain. Government strategic initiatives to digitize production procedures throughout several industries have strengthened global industrial leaders such as Germany, the United States, France, and Japan. Since the inception of industrial revolutions, humanity has persistently progressed, advancing from IR 1.0 to IR 5.0 (Musarat et al., 2023).

5.8 Essential Attributes of Construction 5.0

Cooperative robotics and AI robotics have traditionally been employed for repetitive jobs, but in Construction 5.0, cobots operate alongside people. These robots are meant to help rather than supplant human labour, facilitating activities such as lifting large goods or executing precision work. AI significantly aids project managers in decision-making, risk prediction, resource optimization, Intelligent Infrastructure, and IoT. The IoT facilitates instantaneous communication among machines, materials, and personnel at construction sites. Embedded sensors in equipment and structures deliver

TABLE 5.4
Construction 5.0 Attributes

Attribute	Description	Technologies	Impact
Collaborative robotics	Robots assist humans in tasks (e.g., heavy lifting, precision work)	Cobots, AI-driven systems	Enhanced safety, reduced labour strain
IoT & smart infrastructure	Real-time data exchange between machines, materials, and workers	Embedded sensors, cloud platforms	Waste reduction, hazard prevention
Sustainability focus	Green materials, renewable energy integration, waste minimization	AI, BIM, prefabrication	Lower carbon footprint, energy efficiency
AR/VR integration	3D models overlaid on physical sites; immersive project previews	AR headsets, VR simulations	Fewer design errors, improved stakeholder buy-in
Worker safety	Wearables monitor fatigue/stress; exoskeletons reduce physical strain	Smart helmets, IoT wearables	Reduced accidents, long-term health benefits

immediate data on project advancement, worker safety, and material consumption. This integrated ecosystem mitigates waste, prevents project delays, and enhances safety by identifying possible dangers before their escalation. Table 5.4 elaborates on some attributes of Construction 5.0. It also shows the technologies that have been used and their impacts.

5.8.1 Sustainability and Its Ecological Implications

Construction 5.0 always prioritizes sustainability as a response to global threatening environment issues. Green construction practices, the integration of renewable energy, and the use of sustainable materials are essential. By employing AI and IoT, construction teams can evaluate the environmental impacts of their projects in real time, improving energy efficiency and reducing waste. The use of prefabricated materials, energy-efficient designs, and environmentally sustainable technology connects construction with global sustainability goals.

5.8.2 Augmented and Virtual Reality

With the application of AR, workers are enabled to see three-dimensional models of the real construction site, increasing the precision in real time. While VR is mostly utilized to give simulated views of construction sites, particularly for presentations, safety training provides a hand of stakeholders to make any changes in the project before the actual execution and that is why these two are being frequently adopted in the construction industry.

5.8.3 Worker Safety

The primary focus of Construction 5.0 is on human safety, specially designed for a safe environment of construction workers, and inclusion of smart PEEs like smart helmets, exoskeleton technology, real-time unsafe behaviour detection, and unsafe object detection. All these efforts are carried out to enhance the workers' safety and their efficiency. These devices can evaluate fatigue, stress levels, and physical exertion, providing real-time alerts to prevent accidents and injuries. The human-centred approach ensures that work settings are designed to enhance employee comfort and support long-term health.

5.8.4 The Function of BIM in Sustainable Construction

The core principle of sustainability in construction is to adopt practices and employ resources in building processes that reduce environmental effect. Sustainable building entails monitoring any activities or processes that hurt, deplete, or degrade the surrounding natural environment and life at the construction site. Sustainability in the built environment refers to the design of edifices intended to minimize environmental impact and enhancing socio-economic well-being. Sustainable buildings are constructed to effectively address all relevant issues and implement methods that guarantee a substantial positive effect on the environment. This considers energy-efficient methods, minimizing worksite pollution, lowering carbon emissions, regulating waste output, and investing in green buildings and modular construction using BIM and related technology. Sustainable building design emphasizes the development of construction models that minimize environmental effects. This chiefly entails enhancing energy efficiency, refining resource procurement and utilization, and minimizing waste. Architects and design engineers integrate green design concepts, explore various sustainable construction options, and perform viability and constructability assessments to achieve optimal sustainable value in building. This exercise requires careful planning and precise execution to achieve an efficient and viable building design for long-term sustainability. This requires the project team to consider multiple aspects of construction that ensure sustainability. This includes conceptualization, modelling, site selection, energy planning, material procurement, and waste management. The team must investigate the principles of green engineering to promote planning and sustainable practices in building design and construction. This venue serves as a platform for project planners and specialists to develop sustainable building designs that enhance environmental benefits and fulfil comprehensive construction objectives.

Example: *Refurbishment and Sustainability:* The Faro Building in Madrid is the outcome of an extensive renovation of a 1993 edifice, finalized by late 2024. The renovation of the Faro Building aimed to 'renovate the existing structure while honouring the rhythm and urban landscape' and to 'establish a new architectural standard'. The outcome is a 14-storey edifice encompassing about 13,600 m^2, enhanced by a novel external façade with aluminium louvres to maximize solar regulation.

Performs as a central repository for all stakeholders' digital representations of sustainable project physical and functional data, allowing for thorough building life cycle monitoring. It has made it possible to digitally monitor every stage of a building's life cycle, from design to construction to operation, management, and maintenance, which has greatly improved project efficiency and cost control and reduced associated risks [5]. Using BIM can help with sustainable project schedule planning, as well as proactively determining the necessary building materials, equipment, and financial resources [6]. Real-time work monitoring, cost estimation, discrepancy detection, quality assessment, documentation of product defects, and quick project restoration are all possible with the help of it. Using BIM, green building designers, builders, and operators can improve sustainable building design, construction, and management [7]. Energy consumption, carbon emissions, and ventilation assessments are just a few of the green building functions that may be examined using this analytical tool. The goal is to make the structures more sustainable.

While BIM has broad applicability, it is best suited for green building projects and other uses that necessitate information on energy efficiency and environmental friendliness. This is a strategy for implementing enhancements that promote long-term environmental sustainability. Prior study indicates that the utilization of BIM technology significantly contributes to waste reduction, a crucial aspect of sustainable projects. Curry et al. (2013) proposed that the integration of BIM and LCA was predicated on a life cycle cost evaluation. Bynum et al. (2013) examined the applications of sustainable design across many situations, including those utilized by architects and developers, before analysing the perceptions associated with the implementation of BIM, particularly in the context of sustainable construction. Kota et al. (2014) examined the application of BIM for quantifying daylighting in sustainable buildings using analysis and simulations of daylighting models.

Sustainable growth enhances all facets of construction projects when integrating BIM. This entails the formulation and implementation of enduring tactics, alongside contemporary, state-of-the-art technology. Sustainable construction initiatives, as a sophisticated category of architecture, are garnering global interest. The researchers at Osaka University in Japan initially adopted the 'sustainable architecture' idea at their research centre. The structures are encircled by a verdant yard and a pristine wall, significantly reducing their energy expenses (Qian et al., 2019). Consequently, builders, architects, and designers are investigating innovative building design models that accommodate both individuals and the environment throughout the phases of planning, construction, use, and maintenance as extensively as feasible. These sustainable practices and developments will yield substantial revenues due to diminished carbon dioxide (CO_2) emissions and enhanced energy efficiency while also being more environmentally sustainable. It is recognized for enhancing resource management and safeguarding worker safety throughout construction, leading to reduced waste and diminished exposure to hazards.

The implementation of BIM plays a key role in sustainable design efforts. BIM 3D modelling facilitates the visualization and analysis of diverse environmental parameters associated with the project's constructed area, landscape, and topographical context. Numerous BIM software and information management tools, including Autodesk Revit, ArchiCAD, and Tekla Structures, support thorough environmental impact assessments and quality control planning. These tools enable the identification of relevant sustainable construction alternatives and facilitate informed decision-making aligned with specific project needs.

The introduction of BIM in the design enables construction stakeholders to investigate ideas and develop designs that enhance energy efficiency, minimize waste, and mitigate environmental effects. Recent years have seen significant advancements in the construction sector, transforming project delivery, efficiency, sustainability, and overall productivity [1]. BIM is an IT-based methodology that involves the use and management of a complete digital representation of all building information throughout the various phases of the project life cycle, serving as a data repository. It is a recent innovation that has become an essential instrument for sustainable construction and development, acknowledged for its prospective benefits. The integration of BIM in construction projects yields economic and environmental advantages, improves product quality, and promotes sustainable architectural designs.

The utilization of BIM in sustainable architecture is illustrated by the Shanghai Centre, which realized a 40% reduction in energy consumption through BIM-integrated sustainability strategies (Zhang et al., 2019). The Pearl River Tower employed BIM to decrease energy expenses by 30% and enhance management efficiency (Zhang et al., 2019). The integration of BIM provides numerous advantages for sustainable growth in construction projects. Eco-friendly construction approaches are gaining increasing recognition worldwide. This approach necessitates the formulation and implementation of long-term strategies, alongside the integration of advanced technologies, to achieve its objectives. Sustainable building aims to enhance the social and economic welfare of communities while fostering environmental preservation through diverse construction practices. Table 5.5 represents the strategies adopted for BIM, along with tools and techniques that have been used to get the output.

5.9 BIM and Renewable Energy

A BIM model that simulates actual building design and construction can provide users with appropriate information. BIM is a valuable tool for sustainable integrated design, promoting a new approach to sustainable architecture

TABLE 5.5
BIM Sustainable Strategies

Strategy	Description	Tools/Technologies
Virtual energy simulation	Predicting energy needs and optimizing efficiency	Autodesk Revit, EnergyPlus
Carbon footprint mitigation	Selecting low-impact materials, renewable energy integration	Tally (LCA software), GIS
Logistics optimization	Route planning for reduced fuel consumption	4D/5D BIM, IoT-enabled supply chains
Deconstruction planning	Designing for future disassembly and material reuse	Dynamo, Tekla structures
Indoor air quality	HVAC system optimization for ventilation and filtration	Computational fluid dynamics (CFD) tools

that prioritizes low environmental impact. BIM technology is gaining popularity in building design, construction, and management, and has piqued attention in the hydropower business due to its adaptability to any project. To effectively integrate renewable energy systems with BIM, they must be represented inside a BIM framework. Once simulated, it may be seamlessly integrated into building designs or utilized independently as a stand-alone energy system. An example of the former is solar panels that can be developed and incorporated into a BIM object library for straightforward reuse in building design. A complete photovoltaic system can be modelled in BIM software and installed on-site to supply power to an adjacent building.

5.9.1 Exploring Clean Energy Options with BIM

BIM serves as a robust framework for assessing renewable energy alternatives. The evaluation of a location for solar panels or wind turbines involves an analysis of factors including solar irradiance, wind speed, and shading effects. This analysis facilitates the selection and implementation of appropriate renewable energy solutions, thereby reducing the building's dependence on conventional energy sources and lowering its carbon footprint. The implementation of BIM enables architects and engineers to create structures that are aesthetically pleasing, functionally efficient, and environmentally sustainable. Utilizing BIM to enhance natural light, choosing sustainable materials, and investigating renewable energy alternatives fosters a more sustainable future for the construction sector.

As we approach the promising environment of 2025, architects and engineers must embrace the diverse functionalities of BIM that surpass conventional design practices. Pinnacle Infotech employs a forward-thinking strategy, effectively integrating BIM to ensure that each project transcends mere construction, serving as a testament to sustainable and innovative design. Collaborate with Pinnacle Infotech to initiate a path towards a more sustainable future in building by leveraging BIM technology, which not only constructs edifices, but also fosters a sustainable legacy. BIM enhances sustainable construction using several methodologies.

5.9.2 Enhanced Design Evaluation

Utilizing BIM, building modelers may visualize and simulate designs to gain accurate insights into the project's future progression and outcomes. This assists AEC companies and experts in evaluating construction needs and gaining profound insights into the most effective sustainable design alternatives. This greatly assists construction projects in achieving optimal space use, constructing energy-efficient structures, and maintaining environment-friendly designs to yield sustainable outcomes.

5.9.3 Virtual Energy Simulation

BIM allows construction professionals to assess the energy requirements of a building project and develop energy-efficient models, resulting in a detailed framework for energy consumption and power utilization. This enables construction teams to assess and create proactive virtual energy models that adhere to sustainable construction principles. This facilitates the coordination of various stakeholders and resources for implementing energy-efficient building designs and developing a unified construction plan that promotes sustainability from the outset.

5.9.4 Life Cycle Assessment

BIM-based construction allows teams to perform a thorough assessment of the building structure and its surrounding environment during the project's duration. Architects and engineers can evaluate and measure the environmental impact of construction materials, energy usage, emission levels, and compliance with local requirements using life cycle analysis. This allows construction project managers to execute effective project planning while maintaining adherence to environmental equity and ecological stability throughout the project's duration.

5.9.5 Effective Resource Management

BIM enables the construction industry to utilize advanced technologies and integrated construction methodologies. Teams utilizing the BIM approach can formulate strategies and practices that enhance the efficiency and systematic nature of construction operations. This results in enhanced resource management, facilitating regulated utilization of project resources while reducing idle time and minimizing fuel and equipment waste. This substantially alleviates strain and affects the environment.

5.9.6 Mitigation of Carbon Footprint

Building engineers and architects utilize BIM to adopt various tactics aimed at reducing the carbon footprint through conscientious construction practices. This encompasses various tactics, including the selection of materials with low environmental impact, the optimization of building systems for energy efficiency, and the integration of renewable energy sources into the projects. Successful applications of BIM employed to reduce carbon footprint on the environment encompass the reduction of greenhouse gas emissions in the construction activities of the ZEB Pilot House in Norway and the BRE Innovation Park in the UK.

5.9.7 Eco-Friendly Construction Practices

BIM offers advantages in the development of design structures and information models for buildings, facilitating the effective application of sustainable practices to mitigate significant construction failures and breakdowns. This enables project teams to effectively design and structure workflows that enhance functional planning and job management, thereby significantly reducing operational control and maintenance costs. This is accomplished by simulating building systems in the preliminary construction phase through BIM-based design and fostering sustainable building operations.

5.9.8 Integration of Circular Economy

BIM significantly advances the development of circular economic principles. BIM-based construction enables project teams to adopt strategies for the reuse, recycling, and repurposing of materials throughout the building's life cycle. This sharply contrasts with conventional construction methods that employ materials and resources without regard for their negative environmental impacts. This deliberate decision goes beyond mere usage, emphasizing the enhancement of resource longevity and value, thereby minimizing or eliminating environmental impact.

5.9.9 Eco-Friendly Renovation

One of the other key elements in BIM implementation is sustainable retrofitting, which allows proprietors of antiquated structures to develop digital models for system modernization, improving energy efficiency, and fostering sustainability. This technique utilizes diverse digital tools to identify structural vulnerabilities, outdated systems, and operational hazards via invasive project inspections, employing a shared data environment and cross-structural system information. This technique involves the modernization of outdated structures with energy-efficient windows, HVAC systems, electrical equipment, and plumbing systems.

5.9.10 Intelligent Building Systems

BIM enables engineers to investigate and execute diverse creative options for intelligent building system designs, facilitating highly efficient sustainable implementations for lighting, plumbing, HVAC, and security. This results in an efficient and economical operational system that provides significant value and savings for property owners. Smart building initiatives incorporate capabilities for predictive maintenance and compliance monitoring to ensure optimal performance and enhanced security.

5.9.11 Enhanced Atmospheric Purity

HVAC systems are essential for establishing indoor air quality. Utilizing BIM technology, mechanical engineers may efficiently plan and design HVAC systems to provide optimal airflow, ventilation, filtration, and regulation within the building. BIM enables HVAC design professionals to examine and evaluate the requirements for power equipment, load capacity, and energy flow within a building, ensuring effective management of operational costs and air quality.

5.9.12 Design and Deconstruction

Throughout the course of the project, improved construction planning is guaranteed by valuation engineering in BIM. Project planners, architects, engineers, and other stakeholders can communicate and anticipate future needs with the help of advanced BIM technology, which makes it easier to scale the project strategically through building demolition. The team can confidently explore design options and pursue future projects with efficiency, thanks to this early preparation for either disassembling various building components or repurposing and modernizing the entire structure. This comprehensive BIM-driven approach mitigates excessive resource utilization and waste that could arise from conventional construction methods.

5.9.13 Sustainable Transportation

BIM improves the organization of construction logistics through the optimization of delivery routes and schedules. This markedly reduces resource consumption and fuel usage, resulting in lower carbon emissions and enhanced economic advantages. The incorporation of modern technology and value-driven approaches in 4D/5D construction simulation and phasing improves the efficiency, modularity, and scalability of transportation and logistics operations. This enables efficient route optimization, enhanced resource coordination, and reduced vehicle and equipment idling, resulting in environmentally sustainable transportation.

5.9.14 Site Evaluation and Selection

BIM enables thorough examination of construction sites and surrounding landscapes for site selection using geographic information systems (GIS) technology. This method facilitates a thorough evaluation of several environmental factors, including terrain, climatic conditions, and the area's natural ecosystem. This allows project teams to identify potential environmental threats and proactively eliminate anticipated risks, thus ensuring optimal advancement towards ecological resilience.

5.9.15 Decreased Construction Duration

BIM enhances planning, coordination, and communication among project stakeholders throughout various disciplinary processes and development phases. The construction team enhances time management through the utilization of detailed 3D models, comprehensive work schedules, organized site planning, accurate structural data, and a complete operational blueprint. This enhances processes and markedly decreases overall turnaround time, thereby facilitating sustainable construction.

5.9.16 Environmental Certifications

BIM is crucial for achieving green building certifications, including Leadership in Energy and Environmental Design (LEED) and Building Research Establishment Environmental Assessment Method (BREEAM). BIM enables building engineers to integrate sustainability indicators into the design, thereby improving various environmental factors associated with the project. The contraction models assess and validate various ecological parameters, including energy efficiency, water usage, material type, and indoor environmental quality, to support the team in meeting sustainability criteria for environmentally responsible designs.

5.9.17 Minimized Rework

AECO experts engaged in extensive building projects must maintain controlled and streamlined processes to prevent misunderstandings and discrepancies that typically arise in prolonged and large-scale endeavours. BIM serves as the course curator, enabling project professionals to develop comprehensive plans and clash-coordinated structural layouts to prevent potential collisions and gaps in construction. This aids in reducing the likelihood of redundancy, inefficiency, and rework, hence expediting the execution process. Table 5.6 shows the role of BIM in sustainable construction in the form of applications, along with the methods it followed, and examples.

TABLE 5.6
Role of BIM in Sustainable Construction

Application	Method	Outcome/Example
Design optimization	3D modelling for energy efficiency, material use, and site selection	Faro Building (Madrid): reduced energy use via BIM-driven façade design
Life cycle management	Digital oversight from design to maintenance	Shanghai Tower: 40% energy savings
Renewable energy integration	Simulating solar/wind systems in BIM models	Hydropower plants: optimized turbine placement
Waste reduction	Clash detection, material quantification	ZEB Pilot House (Norway): lower emissions
Certification support	Tracking LEED/BREEAM metrics (energy, water, materials)	Pearl River Tower: 30% lower energy costs
Circular economy	Reuse/recycling planning through BIM databases	BRE Innovation Park (UK): material repurposing

5.10 Industry 5.0 Technologies

Industry 5.0 uses those technologies that have been introduced by Industry 4.0 but have got the influence of human collaboration with these techniques, which include cobots, blockchain, unmanned aerial vehicle (UAV), AI, and DT. Figure 5.3 represents the types of technologies that are being utilized in this manner.

Various technologies have been involved in creation of Industry 5.0 with BIM. Some of them are DT, AI, IoT sensors, blockchain, and AR/VR. Table 5.7 shows the technologies along with their role in BIM and industry applications.

5.10.1 Cobots/Autonomous Robots

Industry 5.0, often known as the Fifth Industrial Revolution, aims to improve industrial processes by reintroducing human technicians. The combination of human and technological intelligence will be a potent force. Collaboration between humans and autonomous machines will be the primary focus of Industry 5.0, in contrast to the automation that dominated Industry 4.0. The autonomous workforce will possess the capability to discern and comprehend human intentions and desires. Humans will collaborate with robots without apprehension, knowing that their robotic counterparts fully comprehend them and can engage in effective teamwork. The results of this productive partnership will be an efficient production process with enhanced value, robust trusted autonomy, and reduced waste and costs.

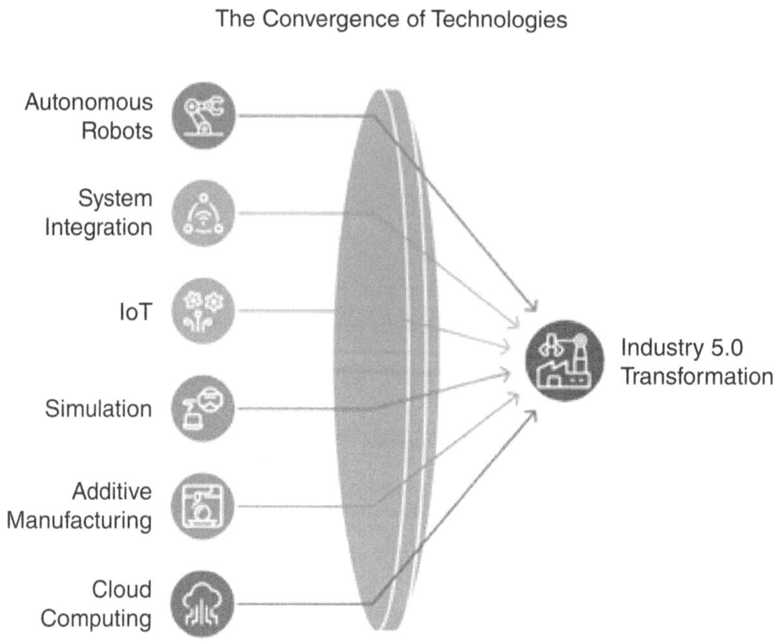

FIGURE 5.3
Technologies in Industry 5.0.

The forthcoming industrial revolution will introduce cobots, which possess pre-existing knowledge or the ability to rapidly acquire new skills, so imparting a human-like quality to robotic manufacturing processes. These cobots will ensure compliance with safety and risk criteria by being cognizant of human presence. They could perceive, comprehend, and relate to human beings as well as human objectives and expectations. More human-centred production environments are the goal of Industry 5.0; thus, Figure 5.4 is

TABLE 5.7
Integration of BIM with Other Technologies

Technology	Role in BIM	Application Example
DTs	Monitoring of assets in real time	Modelling wind farm efficiency for servicing
AI/ML	Predictions of energy usage	Improving the HVAC system in buildings
IoT sensors	Tracking of material and site safety	Construction worker PPEs detection and tracking
Blockchain	Better flow of supply chain management	Tracking sustainable material sourcing
AR/VR	Immersive design reviews	Virtual tours for stakeholder endorsements

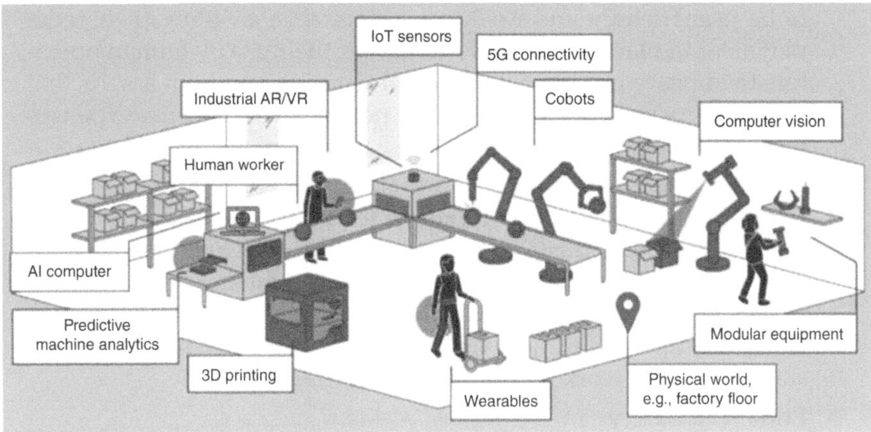

FIGURE 5.4
Human-centric workplaces in which cobots enhance the talents of human workers (Rahman et al., 2024).

shown in this study to depict a typical human-centric setting wherein cobots enhance the capabilities of human workers rather than supplant them.

The significance of cobots and various human–machine interaction technologies and solutions in Industry 5.0 contexts is markedly enhanced in comparison to Industry 4.0. Let us examine the principal distinctions between cobots and conventional industrial robots.

- **Human–Machine Interaction**
  While industrial robots are generally engineered for autonomous operation or integration within a broader machine network, cobots are specifically built for direct interaction with human operators.
- **Significant Adaptability**
  Industrial robots, as advocated by Industry 4.0 principles, are often engineered as specialized instruments for executing one or several specific tasks. Such machines are either impossible or exceedingly challenging to interpret without substantial modifications to their coding and engineering. Cobots are engineered to be versatile and user-friendly devices suitable for a variety of applications.
- **Reduced Dimensions and Diminished Expenses**
  Cobots are significantly smaller than conventional Industry 4.0 industrial robots. This technology is far less expensive to produce, hence enhancing accessibility for organizations and businesses with limited resources for huge industrial machinery.
- **Intuitive and User-Friendly**
  Cobots are transitioning from the intricate complexities characteristic of traditional industrial automation systems. Cobots are designed

to be user-friendly and easily programmable, enabling their redeployment in manufacturing settings and their use by human operators to improve productivity. The Industry 5.0 paradigm asserts that collaborative robotic systems must not involve excessive expenses for installation, programming, and calibration customized to individual user needs.

- **Execution to Do a Broader Range of Jobs**
  In contrast to industrial robots, which are predominantly engineered for heavy manufacturing, cobots are intended to execute a variety of light tasks, including diverse assembly line processes, pick and place operations, material handling, and other functions that support human workers without substituting them.
- **Enhanced Safety**
  Cobots are increasingly employed to manage operations that pose the greatest risk of accidents and injury to human workers. Utilizing them would enhance safety, diminish accidents, and foster improved working circumstances for human personnel.

Example (1): The Hilti Jaibot is a collaborative robot designed for construction applications. This cordless robot, affixed to a mobile platform for enhanced mobility across the construction site, utilizes digital blueprints to delineate and perforate holes. It is also engineered for user-friendliness, especially for individuals lacking robotics knowledge, as shown in Figure 5.5.

FIGURE 5.5
Drilling cobot (Bird, 2023).

FIGURE 5.6
Drywall cobot (Bird, 2023).

Example (2): In March, Hilti, headquartered in Liechtenstein, formed a strategic alliance with Canvas, based in San Francisco. This construction robot firm has created a cobot-driven drywalling solution that utilizes AI and ML, allowing the system to conform to the precise parameters of each job. An image of this cobot can be seen in Figure 5.6.

5.10.2 Blockchain Technology

Manufacturers are presently collaborating on blockchain implementations that could facilitate the optimization of operational processes, enhance transparency in supply chain operations, and effectively manage investments through technology. Blockchain technology has the potential to revolutionize the design, production, and scaling of product lines in manufacturing organizations. Moreover, it is transforming corporate interactions by facilitating collaboration among rival enterprises within shared spaces. Manufacturers still possess significant potential to become innovators in blockchain technology. The sector ranks second to finance regarding perceived technological leadership. Blockchain has the potential to enhance efficiency and expedite operations for future product finance companies. Distributor results and payments can be directly obtained from authenticated blockchain transactions, obviating the need for human intervention and testing. The amalgamation of blockchain technology with cloud computing apps and sensors facilitates exceptional access, availability, and enhanced efficiency throughout the whole value chain. The company's transportation and logistics optimize the synergistic strengths of blockchain and the IoT. Intelligent contracts provide immediate pickups, declines, or business agreements concerning physical products, attributes, or transactions.

5.10.2.1 BIM and Blockchain Technology

The combination of BIM with blockchain technology is revolutionizing the AEC industries by improving transparency, security, and efficiency. BIM, a system for creating and managing data-intensive digital models of structures, seamlessly combines with blockchain, a decentralized ledger technology that ensures permanent and transparent records. They collaboratively address enduring issues like data tampering, supply chain transparency, and contractual disputes.

An essential application relates to permanent information source. The unchangeable ledger of blockchain can securely store BIM data, including design iterations, approvals, and material specifications. Projects requiring ethical sourcing such as the validation of conflict-free minerals or sustainable timber can incorporate blockchain records with BIM models, enabling stakeholders to trace products from origin to installation. Skanska, a prominent construction firm, demonstrates this integration by employing blockchain to verify concrete batch certificates within BIM procedures, thus ensuring compliance with safety and environmental requirements.

Smart contracts, independent agreements recorded on blockchain, enhance BIM processes. These contracts automate actions such as payments or authorization following the achievement of specified project milestones tracked in BIM. The UK's Centre for Digital Built Britain (CDBB) demonstrated this by deploying a blockchain–BIM system for infrastructure projects, thus reducing payment delays and disputes via automated contract execution. Dubai's Blockchain Strategy 2020 mandates the incorporation of blockchain technology with BIM for governmental projects to eliminate corruption and accelerate approvals.

Blockchain enhances the protection of intellectual property (IP) in design. By timestamping and encrypting BIM files on a blockchain, architects and engineers may establish ownership and deter unauthorized utilization of their designs. This is especially beneficial in collaborative projects involving numerous stakeholders contributing to a unified BIM model. Moreover, the integration of blockchain with BIM enhances facility management by preserving immutable maintenance records, hence facilitating audits and compliance verifications. IBM's Maximo employs blockchain technology to monitor asset histories within BIM systems, thereby guaranteeing precise life cycle management.

> *Blockchain guarantees immutable recordings of BIM data; yet excessive transparency may pose challenges to privacy considerations. Should industries implement ethical frameworks to regulate data-sharing permissions inside blockchain-BIM ecosystems? In what manner can human oversight reconcile the inflexibility of smart contracts with contextual decision-making?*

5.10.3 Unmanned Aerial Vehicle

Industry 5.0 represents a sophisticated framework aimed at improving industrial intelligence and connectivity through an integrated platform, incorporating technologies such as the Internet IoT and UAVs. In accordance with the Industry 5.0 framework, UAVs are continuously being developed. UAVs are primarily employed in civilian and military contexts; however, the integration of contemporary manufacturing technologies, such as real-time monitoring systems, wireless visibility, and environmental surveillance, merits investigation. UAVs function as IoT devices that facilitate efficient and cost-effective data collection and monitoring, serving as proactive solutions that improve decision-making processes. UAV technology facilitates industrial growth by linking multiple manufacturing lines through long-distance communication channels. The significance of UAV technology is rising, particularly in its application within smart cities and the IoT. UAV technology has the potential to improve the efficiency of manufacturing facilities, thereby contributing to the advancement of Industry 5.0. A UAV operates autonomously, serving as a medium for transmitting sensory data to a cloud server via network technology, employing a Wi-Fi component. This technology is relevant across multiple areas of pervasive computing, such as pollution management, public transportation, and Industry 5.0, among others.

5.10.4 BIM and Augmented/Virtual Reality

The integration of BIM with AR and VR is revolutionizing the design, construction, and maintenance of buildings, connecting digital models with physical realities. BIM provides a detailed three-dimensional database, while AR overlays digital information onto the physical environment, and VR engages users in interactive virtual spaces. Their collective efforts enhance collaboration, precision, and safety throughout the project life cycle. During the design phase, VR enables architects and clients to assess BIM models at full scale, identifying spatial or aesthetic issues prior to the start of construction. Zaha Hadid Architects utilizes VR to provide clients with immersive walkthroughs of their BIM-designed environments, thereby reducing design modifications by allowing stakeholders to experience layouts in real time. This immersive approach improves decision-making and fosters creativity, allowing designers to explore innovative concepts in a safe virtual setting.

AR technology like Microsoft HoloLens superimposes BIM models onto real-world environments, offering real-time assistance to workers. For instance, Trimble's XR10 headset projects BIM data onto machinery, aiding specialists in the precise assembly of complex components with millimetre accuracy. This integration minimizes errors and rework, particularly in prefabrication, where AR-assisted assembly of BIM-modelled modules ensures seamless on-site integration. In heritage restoration initiatives, like the refurbishment of Notre-Dame Cathedral, VR-BIM technologies digitally reproduce historical edifices, enabling designers to assess reconstruction methods without endangering the actual site. The integration of BIM with AR/VR has transformed safety training. Organizations like Bechtel utilize VR simulations based on BIM environments to train staff in perilous scenarios, such as traversing fire exits or managing heavy equipment. These immersive trainings have reduced onsite injuries by 25%, as staff gain practical experience in a controlled virtual environment. Post-construction, AR enhances facility management by allowing technicians to visualize hidden systems (e.g., electrical wiring or plumbing) through BIM data overlaid on the physical infrastructure. Applications like DAQRI enhance maintenance personnel engagement with BIM models via AR glasses, optimizing repairs, and reducing downtime. In real estate, developers utilize VR walkthroughs of BIM models to promote properties prior to construction, providing buyers with a concrete understanding of space and design. This method has demonstrated efficacy in high-rise constructions, as virtual tours enhance pre-sales and bolster investor trust.

AR/VR facilitates widespread access to BIM models; but immersive technology may undermine practical knowledge. Can hybrid training programmes maintain traditional craftsmanship while utilizing AR/VR efficiency? What measures are necessary to avert cognitive overload from real-time BIM data overlays in critical environments?

5.10.5 Digital Twins

A DT refers to a digital representation of a real system or object. Physical entities such as wind farms, factories, jet engines, buildings, and complex systems like smart cities can be digitally represented through DTs. The concept of DT was introduced in 2002; however, it has only become prominent in recent years due to the rise of IoT technologies. The IoT has made digital transformation economically feasible, thereby improving its accessibility and affordability for many businesses. Data from physical entities is conveyed to their digital equivalents for simulation through IoT devices. The digital mapping of real-time items and systems via DTs allows for the analysis and monitoring of their digital counterparts, thereby enabling the prevention of issues prior to their occurrence in the physical domain. The rapid advancement of AI, ML, and big data analytics has enabled digital transformation to reduce maintenance costs and improve system performance.

In Industry 5.0, DTs can provide substantial value for the creation of tailored products, improved company operations, diminished flaws, and the swift emergence of creative business models to generate revenues. The DT can facilitate Industry 5.0 in addressing technological challenges by swiftly recognizing them, recognizing things that can be altered or refurbished based on productivity, enhancing predictive accuracy, forecasting future errors, and mitigating substantial financial losses. This form of intelligent architectural design allows organizations to achieve economic benefits more rapidly and consistently than previously possible. In Industry 5.0, DTs can be utilized to create simulation models and access real-time computational data, enabling corporations to remotely alter and update physical products. In Industry 5.0, digital transformation is employed for customization that enhances the user's experience about their product requirements and purchasing process. The National Aeronautics and Space Administration states that DT seeks to develop virtual twin versions of tangible phenomena. DT has a wide range of industrial applications.

Wu et al. (2022) characterized DT as a novel application that integrates a succession of technologies, encompassing perception, transmission, computation, modelling, and simulation. The architecture comprised of physical, data, model, and functional layers. The physical layer includes both tangible entities, such as the human body, objects, and physical space, and intangible entities, such as business processes. The data layer served as the foundation of the DT-based application, incorporating data collection, transfer, processing, and storage. The model layer served as the foundation for DT-based applications, utilizing modelling and diverse technologies to represent real objects digitally. The functional layer serves as the direct representation of the digital twin. The system archived simulation results and visual services, offering business applications to meet the needs of various application scenarios. Digital transformation has become increasingly prominent in recent years. The advancement of IoT, big data, cloud computing, AI, and other next-generation information technologies has made the implementation of digital transformation achievable. AI can independently perform data preparation, analysis, fusion, and advanced information extraction of twin data, utilizing the intelligent matching optimal algorithm without requiring data specialists. As a result, it generates a range of services. AI technology can significantly improve data value, responsiveness, and service accuracy in digital transformation, thereby empowering various vertical industries. The DT ecosystem consists of the foundational support layer, data interaction layer, model construction and simulation analysis layer, common application layer, and industry application layer. AI technology is predominantly employed in the simulation analysis layer.

5.10.6 BIM and DT

In the construction industry, BIM is seen as a 'single source of truth' for both geometrical and non-geometrical data, whereas DTs are seen as a 'single version of the truth' that incorporates BIM's physical and dynamic representation, particularly with blockchain integration. When it comes to 'converging technologies' in the construction industry, BIM and DT are the two most important digital tools for managing and exchanging information. BIM is the process of creating a digital representation of a building's physical and functional characteristics using specialized software such as Revit and

ArchiCAD, among others. Among the many uses for BIM are design coordination, construction planning, and facilities management. The models incorporate data pertaining to the structure's design, construction, and operation. All the way through a building's lifespan, from design to construction to operation and maintenance, BIM models can be put to good use. A DT is a digital copy of an actual physical item or system, like a building or a piece of machinery. Sensor readings and information from other sources, such as BIM software, are typically used to build a digital twin. Its purpose is to portray the item's or system's current state. An entity or system's performance, operations, and maintenance can be better understood and optimized with the help of a digital twin, which is a living, breathing digital copy of the physical item (Sepasgozar, 2020). The two technologies are enabling the integration of additional digital technologies to provide essential and complementary solutions for the construction industry.

Cutting-edge BIM and Digital Twin Technologies have taken Construction 4.0 to the next level. Project managers and engineers may optimize designs, run scenarios, and continuously monitor project performance with the use of a twin computer, which is a real-time computer replica of a physical structure. Construction decision-making and long-term asset management are both improved by this technology. Digitalization of goods, processes, and materials is at the heart of new business models that will revolutionize the building industry. DT is expected to develop as a successor to BIM, according to a recent survey of vendors in the construction industry. This will help the sector make the change [8]. With the help of digital technologies, the AEC market is expected to swell from $7,188 million in 2020 to $15,842 million by 2028, representing a compound annual growth rate (CAGR) of 10.7% from 2021 to 2028 (Sepasgozar et al., 2023). Table 5.8 shows different authors' attempts to integrate BIM with other technologies.

Digital Twins enable real-time monitoring of assets; nevertheless, the ownership of the produced data is ambiguous. Should global standards for data protection

TABLE 5.8

BIM and Other Technologies Integration

Authors	BIM	GIS	IoT	ML	AR	Reality Capture	Digital Twins
Deng et al., 2021	✓						✓
Wong et al., 2018	✓	✓	✓			✓	
Rausch et al., 2023	✓					✓	✓
Bosch-Sijtsema et al., 2021	✓			✓		✓	✓
Malagnino et al., 2021	✓		✓				
Gheisari and Irizarry, 2016	✓				✓		
Chen et al., 2021	✓		✓		✓		
He et al., 2021	✓						✓

be implemented in BIM workflows? What strategies can be employed to prevent the degradation of human knowledge resulting from overreliance on automated systems?

5.10.7 BIM and AI

Modern building and landscaping project management practices have come a long way, thanks to the use of modern technology. New approaches have been created using a variety of inputs to improve the accuracy and efficiency of building advancement and management on a worldwide scale through BIM and AI. The initial notion of BIM emerged in the 1960s, whereas AI originated in the 1950s; currently, these processes have advanced and are employed in several projects. Deep learning algorithm is one technique in ML; it is capable of handling huge amount of data, and implementation of these methodologies has now become an essential requirement in construction industry (Bassir et al., 2023). Enhancement of sustainable construction methodologies can be possible by the integration of BIM and ML methodologies, where BIM will perform as digital platform to record geometric and physical data of a building. It will also calculate the material and energy usage within a building, and when this information combines with ML capabilities, this data will be examined to find out the hidden issues and predict the future trends to enhance the sustainable design implementation. As an examples, ML has the capability to evaluate the data of energy consumption of a building to predict future environmental impacts. This allows engineers and architects to visualize and optimize key elements such as energy efficiency, HVAC, daylighting, and selection of suitable materials. By incorporating real-time data, AI enhanced BIM capabilities up to five further dimensions named as 4D (time) and 5D (cost). Figure 5.7 demonstrates the utilization of diverse categories of information.

Skanska, a prominent construction business globally, exemplifies the application of ML-driven analytics on BIM models to diminish energy consumption in building operations. By incorporating occupancy data, meteorological patterns, and use trends into their ML models, they optimized building systems in real time, attaining up to a 30% enhancement in energy efficiency. The Singapore Housing and Development Board (HDB) has utilized ML algorithms on BIM data to enhance natural ventilation and decrease cooling loads in public housing projects, resulting in a substantial reduction in carbon emissions.

Furthermore, ML can facilitate the selection of sustainable materials by evaluating lifetime assessment (LCA) data integrated inside BIM frameworks. For instance, it can provide low-carbon alternatives or forecast the environmental impact of various development scenarios. Predictive maintenance, facilitated by the integration of BIM and ML, prolongs the lifespan of building systems and minimizes resource waste. In the EU Horizon 2020

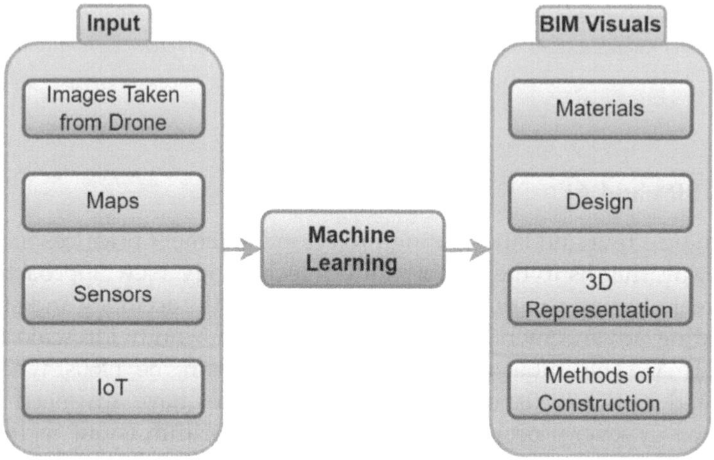

FIGURE 5.7
Information processing by AI for BIM.

BIM-SPEED project, ML algorithms were utilized to examine sensor data from existing buildings to develop retrofitting solutions that reduce energy consumption and enhance occupant comfort. One of the prominent applications of the combination of BIM and ML is structure health monitoring. Structural health monitoring (SHM) is a critical advanced technology. In general, SHM has been designed to enable the transition from post-damage identification to near-real-time or online damage assessment. SHM is a damage detection methodology that enables the long-term observation of a structure through a series of continuous measurement instruments.

Figure 5.8 succinctly illustrates the standard components of a simple harmonic motion system. The process commences with the selection of sensors and their strategic placement on the structure. In order to assess the system's strength, the processing unit stores, processes, and analyses the data transmitted by the data gathering system using a variety of methods and methodologies. The identified damage's severity, location, and its potential for future propagation are considered while making decisions about inspection and maintenance. Modern smart monitoring and decision-making systems are gradually replacing older, more manual damage detection techniques. These examples highlight the revolutionary potential of integrating BIM and ML to attain sustainable objectives. The BIM–ML synergy utilizes data from design to decommissioning, promoting comprehensive life cycle management that advances the construction sector towards net-zero emissions, optimal resource utilization, and robust infrastructure. The Fifth Industrial Revolution emphasizes human-centric and sustainable development, making the integration of intelligent systems with BIM essential for creating greener, smarter built environments.

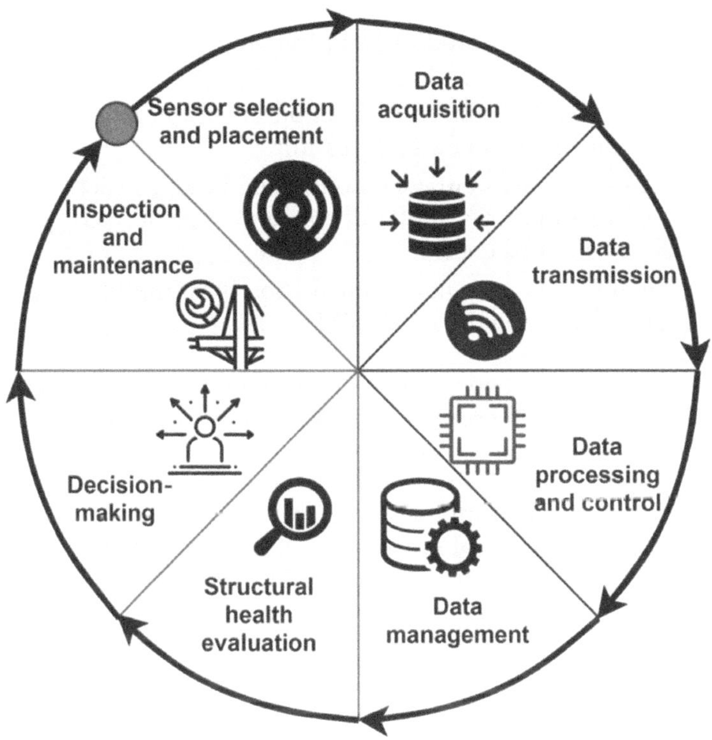

FIGURE 5.8
SHM components (Malekloo et al., 2022).

5.10.7.1 Advantages of Machine Learning in Scan to BIM

The advantages of ML in scan to BIM encompass enhanced accuracy and precision in data processing, expedited workflows, improved clash detection, integrated design automation assistance, increased communication among users, and cost-savings through diminished rework. This facilitates the creation of more dependable and efficient BIM models.

- ML algorithms can generate BIM models with minimal errors by autonomously identifying discrepancies in Point Cloud data.
- Automating the translation of Point Cloud data into BIM elements with ML can substantially decrease the time required and minimize the likelihood of manual errors.
- Feature recognition and element categorization are repetitive operations that can be automated with ML, thereby decreasing the time required.
- ML may identify conflicts among diverse BIM pieces, hence yielding superior quality results.

- ML-generated BIM models exhibit enhanced accuracy and completeness, facilitating improved communication, and coordination among stakeholders.
- ML can yield substantial cost reductions by minimizing rework due to errors and enhancing project planning.
- ML algorithms can accurately model complicated structures due to their ability to process subtle details and sophisticated geometric patterns in scanned data.
- ML can automate as-built documentation by scanning on-site variables and converting them into BIM models.

Example (1): City Reconstruction: The Potential of BIM and AI After Devastation. Kateryna Lopatiuk and Herman Mitish, researchers at the National University of Kyiv, spearhead the 'AI for Cities' initiative, employing computer vision to evaluate drone and satellite imagery for the assessment of damage to structures and infrastructure. Fixed-wing drones outfitted with high-resolution multispectral sensors are utilized to obtain aerial imagery of the impacted cities. A study conducted by McGraw Hill Construction revealed that the implementation of BIM can decrease construction expenses by as much as 20% and shorten construction duration by up to 50%. AI can further augment these advantages by automating tasks, optimizing designs, and enhancing collaboration.

Example (2): Christchurch Earthquake (2011): Following the catastrophic earthquake that struck Christchurch, New Zealand, in 2011, BIM was employed to expedite the city's reconstruction effectively. The city's digital model, encompassing data on edifices, infrastructure, and utilities, allows planners to evaluate the damage, pinpoint priority zones for repair, and synchronize activities among various teams. BIM enhanced resource management and optimized logistics, accelerating the reconstruction process.

BIM users must recognize the significance of employing BIM technology in the development of sustainable structures. For practical applications in building design, it is essential to integrate facilities management specialists into the development process alongside architects and engineers (Manzoor et al., 2021).

5.10.8 BIM and IoT

The term 'IoT' was first coined by the MIT Centre for Automatic Identification Technology in 1999, with its official introduction by the International Telecommunication Union (ITU) occurring in 2005. IoT in the intelligent industry can be utilized for production process control, monitoring of production environments, tracking of manufacturing supply chains, overseeing the entire product life cycle, enhancing safety in production, conserving energy, and reducing emissions. The data generated in these applications is characterized by substantial volume, diversity, velocity, and precision. Nonetheless, BIM has exposed deficiencies, including the mere construction of static models without facilitating dynamic detection, and a lack of adequate interactivity. The digitization of buildings necessitates the optimization of data collection and management; hence, the integration of BIM and IoT, which are two fundamental components of building digitization, has become increasingly significant. The transition from Industry 4.0 to Industry 5.0 signifies a new paradigm in which technology and human-centric solutions integrate. Two essential technologies propelling this revolution in the construction and infrastructure sectors are BIM and the IoT. BIM facilitates the development of digital models of physical environments, whereas IoT links physical equipment to digital networks for instantaneous data transmission. Collectively, they function as essential pillars for more intelligent, secure, and efficient construction methodologies in Industry 5.0.

The Crossrail Project in the UK, a significant railway infrastructure initiative in Europe, employed BIM for its design and construction. This allowed project managers to visualize subterranean tunnels, stations, and utilities to minimize design conflicts and enhancing overall efficiency.

5.10.8.1 IoT

The IoT denotes a network of tangible entities – 'things' – integrated with sensors, software, and various technologies to facilitate connectivity and data exchange with other devices and systems over the Internet.

Principal Attributes of IoT in Construction

- *Real-Time Surveillance:* Monitors equipment, materials, and personnel instantaneously.
- *Predictive Maintenance:* Foresees equipment malfunctions to minimize operational interruptions.
- *Energy Optimization:* Improves energy efficiency using intelligent sensors.
- *Safety Surveillance:* Detects hazardous conditions and transmits notifications immediately.
- *Remote Access:* Facilitates the remote operation and surveillance of machinery.

The Hudson Yards Project in New York City incorporated IoT sensors for real-time surveillance of structural loads and worker safety, hence maintaining elevated levels of safety and operational efficiency. Industry 5.0 emphasizes the utilization of technical innovations while prioritizing human-centred solutions and sustainability. BIM and IoT collaborate to develop intelligent buildings and infrastructure, improving safety, efficiency, and environmental sustainability. The integration of BIM and IoT propels Industry 5.0 by fusing DT with real-time data streams, facilitating more intelligent and adaptive settings. Principal effects encompass. Construction personnel utilize AR glasses to superimpose BIM frameworks with IoT data, such as real-time structural stress indicators, thereby enhancing safety and decision-making processes.

5.11 Principal Contributions

- *Human–Robot Collaboration:* BIM can be combined with robotic apparatus for enhanced precision in building.
- *Augmented Decision-Making:* The IoT delivers real-time data that, when represented in BIM, facilitates more informed decision-making.
- *Improved Worker Safety:* IoT sensors identify dangerous conditions, while BIM simulations forecast threats before their occurrence.
- *Optimization of energy use and material utilization:* This is achieved via digital twin models, enhancing sustainability and efficiency.
- *Digital Twins:* Real-time simulation of physical entities for predictive maintenance and optimization.

5.11.1 Sustainability

Smart buildings such as the Edge in Amsterdam utilize over 30,000 IoT sensors to track occupancy and energy consumption, integrating data into BIM systems to enhance HVAC and lighting efficiency, resulting in a 70% reduction in energy usage. Singapore's Punggol Digital District: BIM-designed sustainable infrastructure incorporates IoT for real-time energy and water management, aiming for net-zero emissions.

5.11.2 Resilience

IoT sensors on structures, such as the Golden Gate Bridge, identify corrosion or stress, while BIM models provide swift repairs, averting failures. Virtual Singapore, a comprehensive 3D city-wide BIM framework combined with IoT data, facilitates flood forecasting and evacuation strategies, hence augmenting disaster resilience.

5.11.3 Applications in Industry

Siemens' Digital Twin Factory integrates BIM-like models with IoT to mimic manufacturing lines, achieving a 50% reduction in downtime. Barcelona's

Smart City Initiative employs BIM for urban planning and IoT technologies to optimize waste management and traffic flow, thereby enhancing the quality of life for residents. BIM–IoT facilities, such as those in the NHS UK, assess equipment efficiency and patient flow, thereby optimizing spatial and energy utilization. BIM and IoT are essential for achieving the vision of Industry 5.0, facilitating the development of more intelligent, sustainable, and resilient systems. By integrating digital and physical domains, they enable individuals to create sustainably, guaranteeing that technology addresses societal requirements. Global case studies illustrate that this synergy is already revolutionizing industries, ranging from building to urban planning, heralding a new era of inclusive advancement.

By 2030, AI-driven building information modelling could diminish project delays by 30%, yet would automation suppress originality in architectural design? Are machines capable of emulating the intuition of experienced engineers?

5.12 Conclusion

The integration of BIM with Industry 5.0 signifies a significant paradigm shift in the AEC industry, especially in promoting renewable energy infrastructure and sustainable development. Technologies like DT, AI, BIM, IoT, and blockchain can be integrated to provide a comprehensive advantage to transform the construction industry by having collaborative effort and efficient operations. These technologies together support transparency and real-time decision-making. This synergy not only expanded the capabilities of BIM, it also focused on safety of construction sites, specially of construction workers as they are the major victims of safety accidents. By this integration, the number of construction accidents can be reduced significantly.

Remarkable advances include usage of BIM in energy simulation to reduce carbon, and the circular economy principles adoption. Examples of these are Pearl Rover Tower and the Faro Building, which show how these integrations are being practised in real world. Technologies in Industry 5.0 have further elevated the efforts by involving cobots which work with human to increase the efficiency and productivity. Use of block makes sure the transparency and the tracking of the flow within supply chain, and AI used for predictive maintenance and informed decision-making. For bridging the gap between human and virtual world, DT provides its role in this regard. Overall, all these technologies have further strengthened the management of the projects and support sustainable development in the construction sector. Nonetheless, issues such as data privacy, ethical governance of automated systems, and the equilibrium between automation and human knowledge necessitate consideration. The industry must implement standardized frameworks to tackle these concerns while promoting innovation. With the increasing worldwide demand for sustainable infrastructure, the integration of BIM and Industry 5.0 will be crucial in attaining net-zero objectives, minimizing waste, and improving resilience. By adopting this convergence, the AEC sector may shift from disjointed processes to cohesive ecosystems, guaranteeing equitable, efficient, and future-oriented construction procedures. The forthcoming journey requires ongoing collaboration, investment in skills, and coordination with global sustainability initiatives to fully harness the potential of these disruptive technologies.

Chapter Review: Chapter 5 – Building Information Modelling and the Fifth Industrial Revolution

Multiple Choice Questions

1. What is the central emphasis of Industry 5.0?
   A. Complete automation of manufacturing facilities
   B. Escalating reliance on fossil fuels
   C. Collaboration between humans and machines about sustainability
2. Which industry is most affected by BIM?
   A. Health care
   B. Construction
   C. Retail
3. What is the meaning of BIM?
   A. BIM
   B. Building information management
   C. Business Integration Model
4. Industry 5.0 underscores:
   A. Substituting humans with robots
   B. Enhancing human skills through technology
   C. Decreasing workforce size
5. Which technology does NOT belong to Industry 5.0?
   A. Digital Twins
   B. Blockchain
   C. Steam engines

6. What is the key objective of integrating BIM with Industry 5.0 technologies in renewable energy projects?
   A. Minimizing human involvement
   B. Optimizing resources and real-time collaboration
   C. Increasing project cost
   D. Reducing automation
7. Which element is emphasized in Industry 5.0 compared to Industry 4.0?
   A. Complete automation
   B. Human-centric technology
   C. Mass production
   D. Fossil fuel dependence
8. In the Architecture, Engineering, and Construction (AEC) sector, Industry 5.0 mainly promotes:
   A. Lower construction costs
   B. Human-centric methodologies, sustainability, and resilience
   C. Increased reliance on manual labour
   D. Single-use materials
9. How does BIM contribute to sustainable construction?
   A. By accelerating project timelines without planning
   B. By minimizing environmental impact through design and life cycle management
   C. By reducing stakeholder communication
   D. By prioritizing low-cost materials
10. What is the role of cobots in Industry 5.0?
   A. Replacing human workers entirely
   B. Enhancing human productivity and safety
   C. Operating autonomously without supervision
   D. Working only in hazardous environments
11. Digital Twins are utilized for:
   A. Real-time asset monitoring
   B. 3D model printing
   C. Social media marketing
12. AI in BIM facilitates:
   A. Predictive maintenance
   B. Enhancing energy efficiency
   C. Accelerating operations

13. Blockchain in building enhances:
    A. Supply chain transparency
    B. Robotic efficiency
    C. Air quality
14. Which project utilized BIM to decrease energy expenses by 30%?
    A. Three Gorges Dam
    B. Pearl River Tower
    C. Eiffel Tower
15. IoT sensors in BIM monitor:
    A. Employee social media engagement
    B. Workplace safety and resource utilization
    C. Equity market fluctuations

Answer Key

Question No.	Correct Option
1	C
2	B
3	A
4	B
5	C
6	B
7	B
8	B

Question No.	Correct Option
9	B
10	B
11	A
12	A
13	A
14	B
15	B

References

Bassir, D., Lodge, H., Chang, H., Majak, J., & Chen, G. (2023). Application of artificial intelligence and machine learning for BIM. *International Journal for Simulation and Multidisciplinary Design Optimization*, *14*, 5.

Bird, B. (2023). Cobots Take on Dull, Dirty & Dangerous Jobs. https://www.ien.com/automation/article/22868200/cobots-take-on-dull-dirty-dangerous-jobs (retrieved 1/5/2025).

Bosch-Sijtsema, P., Claeson-Jonsson, C., Johansson, M., & Roupe, M. (2021). The hype factor of digital technologies in AEC. *Construction Innovation*, *21*(4), 899–916.

Bynum, P., Issa, R. R., & Olbina, S. (2013). Building information modeling in support of sustainable design and construction. *Journal of Construction Engineering and Management, 139*(1), 24–34.

Chen, H., Hou, L., Zhang, G. K., & Moon, S. (2021). Development of BIM, IoT and AR/VR technologies for fire safety and upskilling. *Automation in Construction, 125*, 103631.

Curry, E., O'Donnell, J., Corry, E., Hasan, S., Keane, M., & O'Riain, S. (2013). Linking building data in the cloud: Integrating cross-domain building data using linked data. *Advanced Engineering Informatics, 27*(2), 206–219.

Deng, M., Menassa, C. C., & Kamat, V. R. (2021). From BIM to digital twins: A systematic review of the evolution of intelligent building representations in the AEC-FM industry. *Journal of Information Technology in Construction, 26*, 58–83.

Gheisari, M., & Irizarry, J. (2016). Investigating human and technological requirements for successful implementation of a BIM-based mobile augmented reality environment in facility management practices. *Facilities, 34*(1/2), 69–84.

He, R., Li, M., Gan, V. J., & Ma, J. (2021). BIM-enabled computerized design and digital fabrication of industrialized buildings: A case study. *Journal of Cleaner Production, 278*, 123505.

Kota, S., Haberl, J. S., Clayton, M. J., & Yan, W. (2014). Building information modeling (BIM)-based daylighting simulation and analysis. *Energy and Buildings, 81*, 391–403.

Malagnino, A., Montanaro, T., Lazoi, M., Sergi, I., Corallo, A., & Patrono, L. (2021). Building information modeling and Internet of Things integration for smart and sustainable environments: A review. *Journal of Cleaner Production, 312*, 127716.

Malekloo, A., Ozer, E., AlHamaydeh, M., & Girolami, M. (2022). Machine learning and structural health monitoring overview with emerging technology and high-dimensional data source highlights. *Structural Health Monitoring, 21*(4), 1906–1955.

Manzoor, B., Othman, I., Gardezi, S. S. S., & Harirchian, E. (2021). Strategies for adopting building information modeling (BIM) in sustainable building projects: A case of Malaysia. *Buildings, 11*(6), 249.

Musarat, M., Irfan, M., Alaloul, W., Maqsoom, A., & Ghufran, M. (2023). A review on the way forward in construction through Industrial Revolution 5.0. *Sustainability, 15*(18), 13862.

Qian, Q. K., Fan, K., & Chan, E. H. (2019). Regulatory incentives for green buildings: Gross floor area concessions. In: *Building governance and climate change* (pp. 173–190). Routledge.

Rahman, M. M., Khatun, F., Jahan, I., Devnath, R., & Bhuiyan, M. A.-A. (2024). Cobotics: The evolving roles and prospects of next-generation collaborative robots in Industry 5.0. *Journal of Robotics, 2024*(1), 2918089.

Rausch, C., Lu, R., Talebi, S., & Haas, C. (2023). Deploying 3D scanning based geometric digital twins during fabrication and assembly in offsite manufacturing. *International Journal of Construction Management, 23*(3), 565–578.

Scharmer, V. M., Vernim, S., Horsthofer-Rauch, J., Jordan, P., Maier, M., Paul, M., ... & Zaeh, M. F. (2023). Sustainable manufacturing: A review and framework derivation. Sustainability, 16(1), 119.

Sepasgozar, S. M. (2020). Digital twin and web-based virtual gaming technologies for online education: A case of construction management and engineering. *Applied Sciences, 10*(13), 4678.

Sepasgozar, S. M., Khan, A. A., Smith, K., Romero, J. G., Shen, X., Shirowzhan, S., Li, H., & Tahmasebinia, F. (2023). BIM and digital twin for developing convergence technologies as future of digital construction. *Buildings, 13*(2), 441.

Wong, J. K. W., Ge, J., & He, S. X. (2018). Digitisation in facilities management: A literature review and future research directions. *Automation in Construction, 92*, 312–326.

Wu, Z., Chang, Y., Li, Q., & Cai, R. (2022). A novel method for tunnel digital twin construction and virtual–real fusion application. *Electronics, 11*(9), 1413.

Zhang, L., Chu, Z., He, Q., & Zhai, P. (2019). Investigating the constraints to building information modeling (BIM) applications for sustainable building projects: A case of China. *Sustainability, 11*(7), 1896.

Zhang, L., Chu, Z., & Song, H. (2019). Understanding the relation between BIM application behavior and sustainable construction: A case study in China. *Sustainability, 12*(1), 306.

6

Life Cycle Cost Analysis in Construction Projects and the Future of Intelligent Construction

6.1 Introduction

The use of Life Cycle Cost Analysis (LCCA) in building projects is revolutionizing the assessment of project sustainability and economic viability. As the sector transitions into smart building technologies, comprehending the long-term cost consequences through LCCA is essential for enhancing both financial and environmental outcomes. This chapter analyses the principles and techniques of LCCA in both conventional and contemporary building projects, assessing the implications of these findings for the future of intelligent construction. The 1995 version of the National Institute of Standards and Technology (NIST) Handbook 135 (Dell'Isola, 1997) refers to LCC as 'the total discounted money spent related to the ownership, operation, maintenance, and disposal of a building or building system' over a specified period. LCCA is an economic evaluation technique that determines the total cost of acquiring and sustaining a facility over a certain period. LCCA can be performed on both extensive and minor structures, as well as on specific building systems. Many property owners employ LCCA methods in their decision-making regarding facility development or improvements (Cabeza et al., 2014).

Historically, cost management has been known as a crucial decision-making component for owners and stakeholders during the planning as well as in the execution phase of the project. Cost planning in the planning and design phase, together with monitoring and controlling during the building phase, is an essential management activity that can determine the success of a construction project. Applications of LCCA have provided a dominant advantage in project cost management during the past decade. During the first stage (planning or designing phase of the project). As a result, project managers or engineers have assessed the economic feasibility of several alternatives, highlighting the unique elements of construction, techniques,

DOI: 10.1201/9781003599272-6

and materials. It is an evaluation that compares different choices based on economic performance over a period, including initial or principal and future expenses of a building throughout its life cycle (Lakousha, 2023). As energy efficiency and durability of material are influenced by long-term costs associated with, implementation of LCCA becomes more important for construction projects, especially in the case of building LCCA is a method for assessing the total cost of facility ownership. It includes all costs associated with the acquisition, ownership, and disposal of a building or building system. LCCA is especially advantageous for evaluating project alternatives that fulfil the same performance standards yet differ in initial and operational expenses, aiming to determine the option that maximizes net savings. LCCA will evaluate the cost-effectiveness of incorporating a high-performance heating, ventilation, and air conditioning (HVAC) or glazing system, which may entail elevated initial expenditures but substantially reduce operational and maintenance expenses. LCCA is inadequate for budget allocation.

6.2 History of LCCA

LCCA originated from mid-20th-century initiatives aimed at enhancing long-term cost efficiency, especially in governmental and military undertakings. Figure 6.1 shows the timely events of LCCA. Also, Table 6.1 represents the timeline for LCCA development along with the events and their outcomes.

6.3 Study Period

The study period refers to the duration assessed by the life cycle analysis process. Although the length of the study may not correspond to the actual lifespan of the project, it typically indicates a duration that is less than the project's lifetime. The study period has two primary phases: the construction phase and the service phase. The construction phase encompasses the duration from the initiation of feasibility studies through design and construction, concluding with the commencement of project operations. The service period commences upon the conclusion of the construction phase and terminates at the conclusion of the research or the lifespan. The decision of the research period is contingent upon various factors, including the longevity of the alternatives, the priorities of the owner, and the demands of the market. The owner or project team leader can ascertain it; typically, it ranges from 15 years to 25 years (Ahmed & Tsavdaridis, 2018). Prolonged durations may be relevant, particularly for large-scale undertakings. This study will

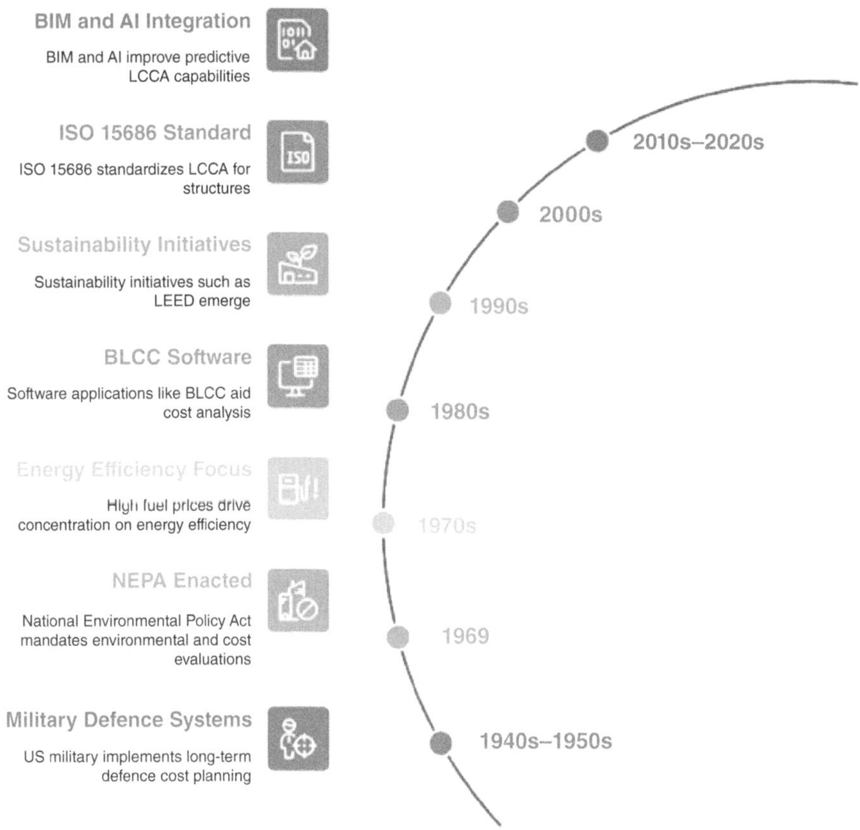

FIGURE 6.1
LCCA development.

standardize the time frame to 30 years, reflecting the extended lifespan of the construction project, to facilitate the comparison of various alternatives. Types of rates in LCCA can be seen in Figure 6.2.

The second crucial component of the LCCA is time. The study period denotes the time frame in which ownership and operational costs will be evaluated. The study duration typically ranges from 20 years to 40 years, depending on the owner's preferences, the stability of the user's programme, and the expected overall longevity of the facility. The research period's duration typically mirrors the expected lifespan of a facility; nevertheless, it is usually shorter than the institution's projected lifespan. The NIST categorizes the research period into two distinct phases: the planning/construction phase and the service phase. The planning and construction phase spans from the initiation of the study until the date the building becomes operational. The lifespan refers to the period during which a building or its elements and components exhibit satisfactory performance, influenced

TABLE 6.1
Timeline for LCCA

Duration	Events	Outcomes
1940s–1950s	The US military implements cost-effectiveness analysis	Emphasized long-term expenditures in procurement (e.g., defence systems)
1960s	The National Environmental Policy Act (1969) requires environmental and cost evaluations	Systematized approaches for assessing the enduring effects of projects
1970s	Energy crises in 1973 and 1979 resulted in a surge in fuel prices	Highlighted operational expenditures, necessitating a concentration on energy efficiency
1980s	Software applications, such as BLCC, facilitate the automation of LCCA computations	Enhanced accessibility for engineers and planners
1990s	Sustainability initiatives, such as LEED, gained momentum	Incorporated LCCA into sustainable building processes
2000s	ISO 15686 standardizes LCCA for structures	Established international standards for uniformity
2010s–2020s	BIM and AI improve predictive LCCA modelling	Facilitated real-time data integration and dynamic forecasting

by several circumstances (Dwaikat & Ali, 2018). Various methodologies exist for assessing service life; ISO 15868-1 and 156862 address the longevity of buildings and their components, offering an approach termed the 'Factors Method' to ascertain the prospective lifespan based on available material and technological data.

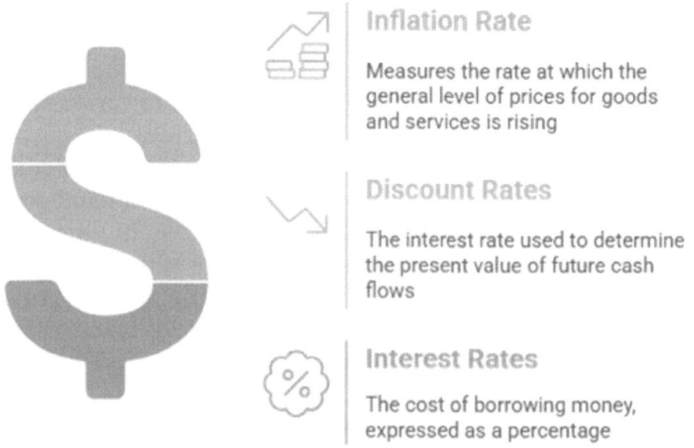

FIGURE 6.2
Rates in LCCA.

The study period refers to the duration assessed by the life cycle analysis process. Although the length of the study may not correspond to the actual lifespan of the project, it typically indicates a duration that is less than the project's lifetime. The study period has two primary phases: the construction phase and the service phase. The construction phase encompasses the duration from the initiation of feasibility studies through design and construction, concluding with the commencement of project operations.

THOUGHT BOX

'Are we planning for today or tomorrow? Traditional cost estimation focuses on short-term objectives. The LCCA encourages us to think in terms of decades, integrating financial planning with long-term performance and sustainability'.

6.4 Inflation Rate

The inflation rate is an economic concept; Craig A. Langston describes it as the rate of growth in the prices of goods and services over time. The primary cause of inflation is the population growth coupled with static resources; to equilibrate supply and demand, prices must be elevated. Inflation is categorized into two types: general inflation, which refers to the overall increase in prices over time, and specific inflation (escalation), which pertains to particular goods or services (Langston, 2013).

6.5 Discount Rates

The discount rate is the interest rate that represents the time worth of money for the investor. For instance, when an investor is presented with the opportunity to receive a sum of money immediately or a greater sum later, the discount represents the adjustment necessary to determine which alternative is most advantageous. There are two categories of discount rates: the first is the actual discount rate, which excludes the inflation rate from its calculation. Conversely, the second type, which exhibits greater accuracy, is the nominal discount rate. The nominal discount rate incorporates the inflation rate in its computations. The first form simplifies computations by omitting the intricacies of determining inflation rates.

6.6 Interest Rates

Interest represents the expense incurred by allocating funds for a specified duration. Consequently, it may refer to the additional funds disbursed to the bank for securing a loan or the monetary gain derived from putting a sum in a bank (Allison, 2012). Interest rate is typically expressed as an annual rate unless specified otherwise; for instance, if a bank's interest rate is 7%, the annual percentage rate of monetary growth is 7%. It emphasizes not only the evident acquisition and construction expenses, but also the operational and maintenance expenditures. In other terms, LCCA examines the whole cost of ownership for a structure or an item. LCCA is a crucial design methodology for managing both initial and subsequent expenses associated with property ownership. LCCA may be executed before design initiatives or at any stage during the design process. Figure 6.3 shows LCCA development and its practices in industry.

It can serve as a helpful tool for assessing the building components that are currently in place. From a whole site complex to a single building unit component, LCCA can assess the overall cost of a variety of projects. Assessing whether an educational capital project is in the best interests of the State of Alaska is the responsibility of the Department of Education and Early Development. Demonstrating that a school district's project proposal benefits not only the district but also the State of Alaska requires the successful implementation of LCCA. LCC denotes the total discounted financial outlay related to the construction, ownership, operation, maintenance, and disposal of a structure or building system over a certain period. Considering this description, the LCC equation may be deconstructed into three variables: the relevant ownership costs, the duration over which these costs are borne, and the discount rate applied to future expenses to align them with present costs.

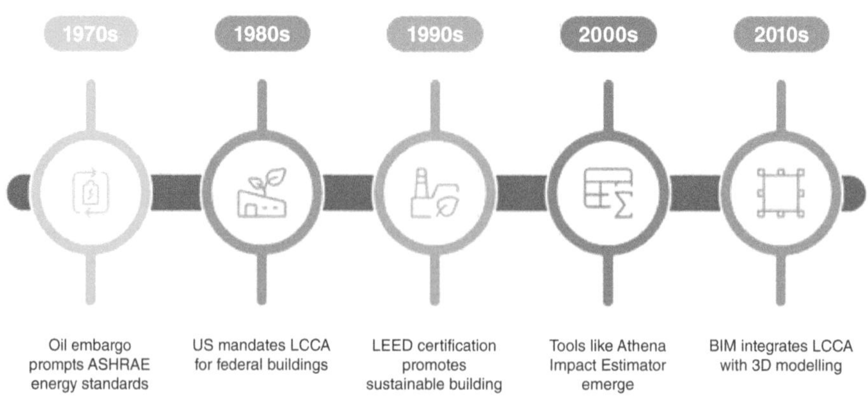

FIGURE 6.3
Evolution of LCCA practices.

TABLE 6.2

Difference between LCCA and Traditional Cost Analysis

Aspect	Traditional Cost Analysis	LCCA
Focus	Rely on initial expenses only	Whole life cycle, design to disposal
Costs	Capital cost only	Operation maintenance and other phases
Time period	Up to construction phase only	Longer duration 20–50 years
Decision	Lowest cost	Balance between short-term and long-term cost
Sustainability	Not considered mostly	Considered

Table 6.2 shows the difference of approach between traditional cost analysis and LCCS for different aspects of projects.

Life cycle costing is a methodology that assists construction industry stakeholders in estimating the total expenditure on an asset over its useful life. The useful life of an asset is the projected duration in years during which it will generate profit. This estimate assists in assessing the duration for which the asset will remain in optimal condition. Financially, it denotes the period during which the assets will yield economic advantages for the business. This procedure covers all expenses related to an asset from purchase to disposal or removal. Unlike traditional cost analysis, which typically emphasizes initial acquisition costs, LCC includes operational, maintenance, and supplementary expenses that will be analysed afterwards. LCCA is the method by which the project team assesses the costs related to a structure, construction material, or equipment across its entire operational existence. All types of costs involved in LCCA are mentioned in Figure 6.4. LCCA is a methodical

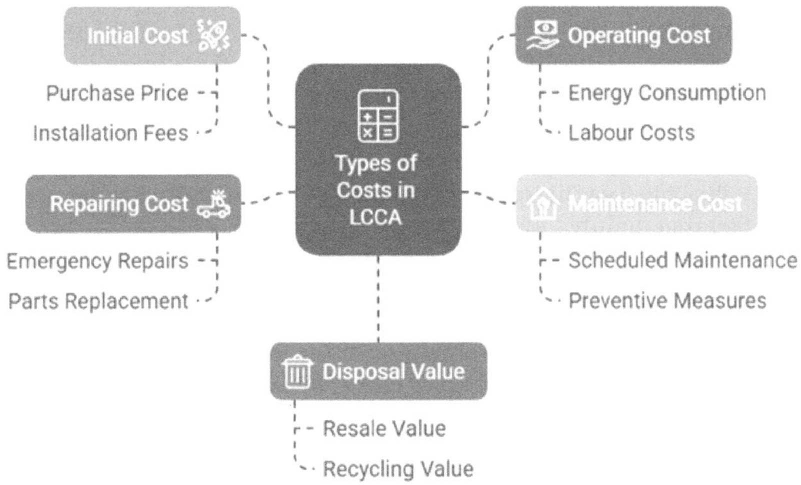

FIGURE 6.4
Types of costs in LCCA.

framework for assessing the comprehensive cost of ownership of construction projects, covering all expenditures from original investment to decommissioning. This analysis aids project managers and stakeholders in making informed decisions by evaluating various project designs and materials in terms of long-term cost efficiency.

6.7 Initial Cost

Primarily the first element of LCCA is cost. In a life cycle of the project two types of cost occur: initial expenses and future expenses. Initial expenses encompass all costs incurred before the occupation of the institution. Future expenses encompass all costs incurred subsequent to the occupation of the institution. Determining the precise expenses of each expense category can be challenging, as nearly all prices are unknown at the time of the LCC study. Nevertheless, by employing rational, consistent, and thoroughly documented assumptions, a valid LCCA can be developed. It is important to acknowledge that not all expense categories pertain to every project. The preparer must include relevant cost categories to ensure an accurate LCC comparison of project options. If expenses in a specific cost category are uniform across all project alternatives, they may be recorded accordingly and excluded from the LCC comparison. The initial expenses encompass the acquisition cost and installation or building fees. These are the initial expenses related to the selected building materials or equipment. These costs significantly influence purchase decisions due to their substantial impact on the project's budget at the time of the decision.

Nevertheless, initial expenses represent merely the surface of a larger issue. Numerous hidden expenses must be considered when assessing the actual cost and worth of an item.

Selecting the choice with the lowest initial cost while disregarding long-term expenses may ultimately prove to be more costly. An object priced at $3,000 with a five-year lifespan will possess less value than a similar one costing $5,000 that endures for ten years. Consequently, exceeding your building project's budget for minimal long-term savings is not advisable. Your project team must diligently comprehend your budget and provide options and explanations that explicitly delineate the whole LCC of the item. This enables you to make a well-informed selection that is optimal for your project. The purpose of a project option's cost analysis (LCCA) is to outline all of the alternative's initial investment costs. Expenses incurred prior to the facility being occupied are referred to as initial investment costs. The entire value of all original costs will be included in the LCCA total. The amount of detail in the project should match the level of detail of these charges. The DEED Cost Model worksheet, construction cost literature, contractor estimates, or expert cost estimating consultants can all be used to calculate construction costs. This encompasses all expenses necessary for the construction or acquisition of the asset, including planning, design, equipment, materials, labour, and overhead expenditure. For instance, preliminary construction expenses can fluctuate considerably depending on the project's intricacy and material selections. Means data indicates that initial construction costs represent between 25% and 35% of the overall LCC.

6.8 Operating Costs

One of the initial post-construction expenses that encounters is the operational cost. This encompasses all needs, from gasoline for operating the equipment to routine cleaning and maintenance. These expenditures are primarily estimated as they tend to vary from year to year or month to month. Nevertheless, these estimations can offer significant insight into your long-term expenses. The subsequent step in finalizing the LCCA of an alternative project is to delineate all prospective operational expenses associated with the alternative. The operational expenses are yearly costs, excluding maintenance and repair expenditures, associated with the facility's operation. Most of these expenses go to building utilities and custodial services. All operational expenditure must be reduced to their current value before being included in the LCCA total. Appendix A enumerates the minimal operational cost categories to be considered in the LCCA. Operational expenses not directly associated with the building should often be omitted from the LCCA. An instance of an expense that needs to be omitted is the expenditure on office supplies. Although it constitutes an annual operational expense, it is unrelated to the building's operation and is instead contingent upon the building user. However, if project alternatives provide varying user requirements, it is vital to incorporate these expenses. An instance of this scenario is the juxtaposition of a year-round educational model with a conventional nine-month academic calendar. The two alternatives may necessitate distinct staffing requirements. Although staffing is not a construction operational expense, it ought to be incorporated in the LCCA to ensure a precise comparison of the alternatives.

More efficient equipment might incur lower operational costs. Nonetheless, they typically possess advanced technology, which is associated with greater cost. In terms of finishing and construction materials, cost-effective alternatives may necessitate more regular maintenance or cleaning, whereas more durable and higher-quality products can endure greater usage.

When evaluating alternatives, do not disregard the operational expenses. Although they may appear trivial, it is important to consider that these expenses will recur throughout the item's existence. They can rapidly accumulate to a substantial amount. Operating costs refer to the daily expenditure necessary for facility operations, encompassing personnel, utilities, and supplies. Maintenance expenditures encompass regular activities necessary to maintain the asset's operational integrity, including inspections, cleaning, and minor repairs. Research by Building Maintenance Management (BMM) indicates that these expenses generally fluctuate between 2% and 4% of the initial building expenditures annually.

6.9 Maintenance Costs

Another element to consider is the maintenance associated with the item. What is the ease of locating qualified maintenance specialists capable of diagnosing and servicing your equipment? Will new parts be readily available, or will they necessitate international procurement or specialized fabrication? What are the characteristics of the warranties? Will the manufacturer provide support for their product, or will you incur substantial costs if the equipment malfunctions? Finding all future maintenance and repair expenses associated with the alternative is the third step in creating the LCCA for a project option. The department's LCCA spreadsheet has combined maintenance and repair expenses to make it easier to use. There is a noticeable difference between the two expenses. The expected cost of maintaining the facility is known as maintenance cost. The cost of sealing the building's roof penetrations and conducting an annual roof inspection are two examples of maintenance expenses. The purpose of this planned activity is to maintain the integrity of the building.

These enquiries can assist in directing a decision-making process regarding particular product. Although we aspire for all equipment to be durable and function flawlessly for years, this is not invariably the reality. If issues arise, cost-effective maintenance can enhance the total worth of the item, even if it is marginally pricier. The third element in preparing the LCCA for a project option is to identify all future maintenance and repair costs related to the alternative. To facilitate usability, maintenance and repair costs have been aggregated in the department's LCCA spreadsheet. A distinct disparity occurs between the two expenditures. Maintenance costs are anticipated expenses associated with the upkeep of the facility. An example of a maintenance expense is the charge for an annual roof inspection and the sealing of the building's roof penetrations. This task is a scheduled endeavour aimed at preserving the building's integrity.

Repair costs are unforeseen expenses necessary to extend the lifespan of a building system without its replacement. An illustration is the restoration of a shattered window. This is an unscheduled occurrence that involves only the replacement of the broken pane, not the entire window unit. Certain maintenance expenses are incurred on an annual basis, while others occur less frequently. Repair costs are inherently unpredictable, making it impossible to estimate when they will arise. For clarity, maintenance and repair expenses should be regarded as annual expenditures. All maintenance and repair expenses must be reduced to their present value before being included in the LCCA total. Throughout a building's lifespan, numerous components may require maintenance or substitution. These expenses can be substantial and should be projected in the LCCA. HVAC systems often necessitate significant repairs or replacements every 15–20 years, with expenditures averaging between $2,500 and $6,000 per unit, as reported by the American Society of Heating, Refrigerating, and Air-Conditioning Engineers (ASHRAE).

6.10 Replacement Costs

No entity remains forever, and ultimately, certain sections of the structure or apparatus will necessitate replacement. All expenses related to the replacement of that item must be included in the LCCA. This may involve the actual removal or uninstallation of the object, the procurement of a new item, and the installation of the new thing. Although one cannot forecast with absolute certainty the market conditions for an object at the conclusion of its useful life, contemplating the potential outcomes is essential. Material shortages may elevate costs beyond inflation, although technological advancements could render it more economical. The frequency and cost of replacement will significantly influence the ultimate value of your project. For instance, it may seem alluring to substitute your carpet with a more economical alternative during the value engineering phase; however, if it necessitates replacement within a few years, the cost savings may not be warranted. The fourth phase in the final stages of the LCCA for a project selection is to define all potential replacement costs associated with the alternative. Replacement costs refer to projected expenses for essential components of building systems necessary for the continuous operation of a facility. All replacement expenses must be reduced to their present value before being included in the LCCA total. Appendix A specifies the fundamental categories of replacement costs to be included in the LCCA. Replacement costs are generally incurred when a building system or component has reached the conclusion of its useful life and requires substitution. An illustration of replacement cost is the substitution of a boiler. The lifespan of a boiler is shorter than that of the facility it operates within. Ultimately, it will deteriorate and necessitate replacement to maintain the facility's operational efficacy.

6.11 Disposal Fee/Residual Value

LCCA evaluates the expenses associated with a product over its full lifespan. This encompasses the scenario where the product attains the conclusion of its functional lifespan and requires disposal. A future expense that requires additional clarification is residual value. The residual value is the net worth of a building at the conclusion of the LCCA study period. In contrast to other prospective expenses, an alternative's residual value may be either positive or negative, representing a cost or a benefit. This fee is often overlooked, as the majority of our everyday disposal can be performed with relative ease and at no charge. Nonetheless, certain construction materials or equipment may result in supplementary costs. The fifth step in finalizing the LCCA for a project option is to determine the residual value of that alternative. The residual value is the net worth of a building or building system at the end of the LCCA study period. This is the only cost category in a LCCA where a negative value, which reduces costs, is acceptable.

As LCC represents a total of expenses, a negative residual value signifies that the building possesses worth after the research period. The value may stem from a freshly replaced roof or the building's superstructure, which could remain functional for an additional 30 years. The residual value, regardless of its cause, constitutes a physical asset of property ownership and must be incorporated into the LCCA. A positive residual value signifies that disposal expenses are linked to the building at the conclusion of the research period. The expenses may be for the remediation of dangerous materials or the demolition of the building. Regardless of the origin, these expenses associated with property ownership must be incorporated into the LCCA.

After the study period a zero residual value indicates that there is no longer any value or cost associated with the building. This rare circumstance occurs when the building's intended use ends concurrently with the end of the study period, the owner is unable to sell the facility, and the owner is free to leave the building without incurring any expenses. When evaluating project options with different life expectancies, the residual value of a building system or facility is very important. The evaluation of two roofing options serves as an example: comparing a composite shingle roof with a metal roof. If your item contains dangerous materials, it may necessitate specific disposal or recycling methods, potentially incurring an additional price. Nonetheless, certain objects may possess a favourable resale value, enabling you to profit by transferring them to a second-hand purchaser or a salvager. The disposal of materials, finishes, and equipment for your new building, although not the primary expense considered, is a significant cost that must be incorporated into the LCCA. Table 6.3 summarizes all the above-discussed costs along with their small description and examples.

TABLE 6.3
Different Costs in LCCA

Cost Component	Description	Examples
Initial costs	Costs spent during the start of a project, such as design, procurement, construction, or execution	Design, land acquisition, construction
Operating costs	Typical costs throughout the operational phase	Energy, water, labour, and utilities
Maintenance	Costs involved in timely repairs and replacements	HVAC servicing and roof repairs
Disposal	Expenses incurred in the safe removal, demolition, recycling, or disposal of the item at the end of its useful life	Waste removal and decommissioning

6.12 LCCA in Building Construction

The building sector implemented LCCA to tackle increasing energy expenses and sustainability requirements, transitioning emphasis from initial costs to long-term value. In the construction industry, life cycle costing is the technique employed to determine the comprehensive cost of a structure, construction materials, or machinery across its entire lifespan. The objective of LCCA is to evaluate the cost-effectiveness of various investment alternatives by considering all direct costs incurred by a decision-maker across the whole economic lifespan of the investment. LCCA has been regarded as a significant analytical method in previous studies and has been extensively utilized in empirical research on building construction. The expenditures encompassed in LCCA pertain to development, construction, agency, occupancy, and environmental factors. The outcomes of LCCA are contingent upon the quantity and precision of its input parameters. The initial phase of LCCA involves the selection of design alternatives grounded in economic factors. Before evaluating environmental and social advantages, any developer must ensure the financial viability of a capital improvement project. Consequently, analysing the operational lifespan of every building system is crucial when considering design alternatives.

For instance, a building's structural system and water heating system will undoubtedly have different lifespans. In a similar vein, the flooring finishes' useful life will probably be much shorter. It is essential to take into account all uncertainties during the execution of LCCA, including the longevity of building systems, prospective operational costs, upkeep, and system replacements. Moreover, the discount and inflation rates are essential components in determining the LCCA of a project. The construction expenses of a building represent just 10–20% of its whole lifetime costs. LCC offers a persuasive viewpoint: construction constitutes only a small portion; the majority of costs, accounting for 80–90%, stem from maintenance and renovations over a building's existence. Similar to how we plan for our financial future through savings accounts and retirement schemes to improve our quality of life, LCC aids businesses and project managers in ensuring the sustainable financial viability of buildings and infrastructures from inception to completion.

THOUGHT BOX

How did crises influence cost analysis? Economic shocks and energy crises have historically driven advancements in cost forecasting. LCCA arose not out of comfort, but from the need to make better decisions in uncertain times.

Does the chosen study period reflect reality or convenience? In LCCA, the length of the study period has a significant impact on the outcomes. Choosing it intelligently while balancing project life and stakeholder goals is critical for effective analysis.

Approaches in LCCA

- *Net Present Value (NPV):* A technique for determining the present value of anticipated cash flows, facilitating the comparison of long-term project expenditures.
- *Internal Rate of Return (IRR):* Assists in assessing the profitability of prospective investments.
- *Discounted Cash Flow (DCF):* Accounts for the time value of money to evaluate prospective costs and benefits.

Example

- *A sustainable office building in Singapore employed LCCA methodologies during its planning phase. The project realized a 20% reduction in LCC over 30 years by maximizing energy-efficient materials and using intelligent energy systems in contrast to traditional designs.*
- *The 'Green Home Initiative' in Denmark, which achieved a 30% reduction in heating expenses through design decisions informed by LCCA.*
- *High-rise buildings in New York City currently utilize LCCA for HVAC systems to forecast long-term savings.*
- *The Millau Viaduct in France employed LCCA to forecast maintenance expenditures throughout its 120-year planned lifespan.*

6.12.1 Reducing Expenditure

To reduce the expenditure of the project, construction managers need to take data-driven decisions as LCCA provides a detailed cost expenditure structure for the whole life of the project, and with the help of this decision, it enables them to make these decisions. Due to this, they can compare different available options, for example, either to do a renovation or demolition of a particular project.

6.12.2 Reducing Carbon Footprint

To find out the cost associated with energy consumption, greenhouse gas emission, and waste production by buildings, engineers can utilize the LCCA technique to determine the cost associated with them. By incorporating the cost into environmental factors, they can make better decisions to reduce them and find a suitable substitute that releases less CO_2. Substituting traditional concrete with low-carbon alternatives might reduce CO_2 emissions by up to 40% by 2030. Implementing renewable energy sources can significantly reduce a building's carbon footprint. Implementing LCCA in construction projects is essential for organizations aiming to attain net-zero CO_2 emissions.

6.12.3 Following Rules and Regulations

In the construction industry, regulatory regulations ensure that projects comply with established quality and safety requirements while preserving economic viability throughout their lifespan. Therefore, the execution of LCC analysis is crucial to evaluate the total expenses of an asset and determine its financial feasibility. Furthermore, adherence to these regulations is crucial for obtaining financing and gaining approval from regulatory bodies. The European Union Directive on Public Procurement (2014/24/EU) mandates that member states assess LCC when evaluating bids for public contracts, encompassing construction projects.

6.12.4 Additional Factors Influencing LCCA

A project represents a significant investment, whether it is new or a renovation. The decisions made during the planning phase can significantly affect both the construction and the budget for years ahead. A comprehensive LCC study of building materials and equipment can facilitate optimal decision-making for both your current and future needs. Although forecasting the future is unattainable, collaborating with a construction firm that can evaluate LCCA through the prism of extensive industry knowledge helps facilitate this process. The time frame from the building's completion to the end of the research project is included in the service period. An Alaskan school project's planning and construction timeline could take many years due to the unpredictable nature of construction finance and the short building season. The department supports the assumption that all initial expenses will be incurred in the base year of the study in order to remove uncertainty over the appropriate length of the planning/construction phase and to expedite the LCC calculation. As a result, the full amount of all original expenses will be documented in the LCCA. The DEED suggests a research duration of 20 years for LCCA. This results from population variations within communities, the dynamic character of educational programmes, the varying lifespan of specific building systems, and the diminished economic impact of expenses incurred beyond two decades.

6.12.5 Lifespan

The lifespan denotes the duration during which a building or its elements and components maintain acceptable performance, varying according to numerous factors (Dwaikat & Ali, 2018). Various methodologies exist for assessing service life; ISO 15868-1 and 156862 address the longevity of buildings and their components, offering an approach termed the 'Factors Method' to ascertain the prospective lifespan based on available material and technological data.

6.12.6 Real Discount Rate

Life Cycle Costing for Design Professionals, 2nd Edition, defines the discount rate as 'the interest rate that signifies the investor's time value of money'. It is third component in the equation. The interest rate is what makes an investor unconcerned about a larger future payment versus an immediate payoff. Real and nominal discount rates are the two categories into which the NIST divides discount rates. The distinction between the two is that the nominal discount rate takes inflation into account, whereas the real discount rate does not. The use of real discount rates simplifies the difficulties of including inflation into the present value formula, not that they ignore it. Equivalent results are obtained when either discount rate is applied in the relevant present value computation. The use of actual discount rates in determining LCC for project alternatives will be the main emphasis of this handbook. LCCA utilizes present-value analysis, requiring the discounting of future costs to their present value using an appropriate discount rate. The choice of discount rate can significantly impact the outcomes of the LCCA. Federal projects often employ discount rates provided by the Office of Management and Budget (OMB), typically ranging from 3% to 7%.

6.12.7 Constant Dollars

Like discount rates, costs can be categorized as real or nominal. Constant dollars are 'dollars with consistent purchasing power associated with a reference year, free from general price inflation or deflation', according to the NIST Handbook 135, 2022 edition. Current dollars are described by the NIST as 'dollars with variable purchasing power, reflecting general price inflation or deflation, in which actual prices are articulated'. Costs must be stated in constant dollars when calculating present value using the real discount rate. Similarly, expenses must be stated in current dollars when utilizing the nominal discount rate for present value computations. Constant dollars are equal to current dollars under the uncommon event of zero inflation, and the real and nominal discount rates are equal.

6.12.8 Present Value

Prior to integrating initial and future expenses, it is necessary to determine the present value of each expense. Current value is 'the time-equivalent value of past, present, or future cash flows at the commencement of the base year', according to the NIST Handbook 135, 2022 edition. To calculate a cost's present value in the base year of the study period, the present value computation uses the discount rate and the cost's occurrence date. Since the majority of beginning costs happen all at once, they are regarded as occurring during the research period's base year. Since their current value is equal to their original cost, it is unnecessary to determine the present value of these early expenses. Time is necessary for calculating the present value of future expenses. The term refers to the time frame between the original expenditures and the subsequent future expenses. In Year 0, the baseline year, initial expenses are incurred at the start of the research period. From Year 1 to the conclusion of the research period, future costs could arise at any point. The equalizer that makes it possible to combine initial and future costs is the present value computation. The present value of future expenses is determined by the discount rate and time. Future expenses will have a lower present value than their true cost because of the positive discount rate (inflation). Equation (6.1) displays the formula for the discount rate:

$$PV = At \times \frac{1}{(1+d)^t} \tag{6.1}$$

Here

PV = Present value

At = Amount of one-time cost at a time t

d = Real discount rate

t = Time (expressed as number of years).

Renovation of an Energy-Efficient Office Building

An office block in downtown Chicago underwent a thorough refurbishment to improve energy efficiency and decrease operational expenses. The renovation encompassed enhancements to the HVAC system, lighting retrofits, and window replacements. The project extended over two years and sought to attain LEED Gold certification. This case study emphasizes essential factors for efficient LCCA, methodologies for cost calculation, and the tools and techniques utilized in the analysis.

6.12.8.1 How Can We Integrate LCCA with Smart Construction?

The Internet of Things (IoT) and real-time data sensors integrated into infrastructures (such as strain gauges in bridges and smart meters in buildings) gather real-time data regarding energy usage, material deterioration, and occupancy trends. An intelligent HVAC system modulates energy consumption according to occupancy data, directly incorporating operational cost reductions into LCCA models.

Effect: Mitigates uncertainty in long-term cost forecasts by substituting assumptions with empirical evidence.

Artificial Intelligence (AI) and Predictive Analytics: Machine learning algorithms examine previous data to anticipate maintenance requirements, energy consumption, and the likelihood of component failures. For instance, AI forecasts the degradation of a roof membrane by analysing meteorological trends, facilitating pre-emptive replacement and reducing life cycle expenses. Platforms such as IBM Maximo and Siemens MindSphere incorporate AI into asset management.

Dynamic LCCA Models: Intelligent construction instruments provide dynamic LCCA models that refresh in real time as new data is received (e.g., variations in material costs and energy rates). A digital twin of a smart building recalibrates its LCCA forecasts regularly, informed by real energy use and tenant activities.

Advantage: Stakeholders can implement mid-project modifications to enhance cost efficiency (e.g., substituting less expensive materials while maintaining durability).

6.13 Synergy between BIM and LCCA

Building Information Modelling (BIM) facilitates the simulation of many design alternatives while enabling immediate calculation of their LCC. Testing insulation materials within BIM indicates that a 10% increased initial expenditure for aerogel insulation results in a 25% reduction in HVAC costs over a 30-year period. BIM software, such as Autodesk Revit, facilitates direct data export to LCCA systems like Tally and ECONO, hence optimizing cost analysis processes.

Case Study 1: Intelligent Healthcare Facility in Stockholm, Sweden

Project Overview: A 20-storey hospital conceived utilizing Building Information Modelling (BIM, Internet of Things (IoT) sensors, and Life Cycle Cost Analysis (LCCA) from the outset.

Innovations

BIM-Driven Design: Enhanced room configurations to diminish construction waste by 15%. *IoT Integration:* Sensors assess air quality and equipment efficiency, reducing operational expenses by 20% each year.

The LCCA outcome indicates that increased initial investments in technology, such as smart lighting, resulted in a 35% decrease in total lifespan costs over a span of 50 years.

Case Study 2: Intelligent Transportation Corridor in Texas, USA

A 100-mile highway using embedded sensors and digital twins for maintenance planning.

Innovations

Models traffic loads and meteorological effects to forecast pavement deterioration. *Predictive Maintenance:* AI orchestrates repairs during periods of low traffic, hence reducing disruptions and expenses. Maintenance budgets decreased by 18% in comparison to conventional reactive methods.

6.13.1 Sustainability and LCCA

- *Circular Economy Integration:* Material Passports: Intelligent construction instruments monitor the provenance, composition, and possibility for reuse of materials (e.g., through blockchain technology). A modular office structure employs QR tags on steel beams to document its life cycle, facilitating future dismantling and resale. Reduces end-of-life expenses by 30–40% via recycling revenue.
- *Adherence to Regulatory Standards:* Tools such as One Click LCA automate carbon cost computations for certifications like as LEED or BREEAM. A net-zero educational institution in California employed LCCA to substantiate expenditures in solar panels, thereby obtaining tax benefits and achieving LEED Platinum certification.
- *Carbon Cost Integration:* Contemporary LCCA models increasingly incorporate carbon taxes and embodied carbon, such as emissions from concrete manufacture. A 2025 EU directive necessitates that LCCA for public projects incorporates Scope 3 emissions, hence promoting the utilization of low-carbon materials.
- *Ethical Considerations:* Overreliance on algorithms may overlook societal repercussions, such job displacement due to automation. Integrating LCCA with Social Life Cycle Assessments (SLCA) is recommended for thorough decision-making.

THOUGHT BOX

Is it more intelligent or more economical? Long-term evaluations of performance frequently fail for buildings designed only based on initial cost. Under LCCA, architects and engineers are enabled to emphasize real value over superficial cost savings.

'In reality, how much are you saving? Long-term expenses could result from short-term savings. By highlighting places where strategic investment generates substantial benefits, LCCA helps prevent false economies'.

What if your structure had the ability to think? AI, sensors, and real-time data turn LCCA from a static blueprint into a dynamic tool that changes as the building it examines does.

6.14 Future Innovations in LCCA

As building and infrastructure projects become more complicated, the evolution of LCCA is not only required, but is also unavoidable. Traditional LCCA procedures, while still essential, must now adapt to a quickly changing technological context. Emerging digital technologies and frameworks are changing the way cost efficiency, sustainability, and long-term resilience are measured. The future of LCCA lies in the incorporation of sophisticated technologies that improve precision, transparency, and adaptability throughout a project's life cycle. Three essential enablers are set to revolutionize LCCA: resilience, blockchain, and generative AI, which can also be seen in Figure 6.5 along with their impacts.

6.14.1 Generative AI for LCCA

Tools such as ChatGPT-4 or Claude may facilitate the automation of report generation, scenario analysis, and stakeholder engagement. An AI assistant composes an LCCA report analysing 50 design options within minutes. Generative AI can significantly improve LCCA through automated cost simulation, modelling design alternatives, and anticipating future scenarios based on large datasets. Powered by AI LCCA, models can use real-time data, historical performance, and predictive analytics to help make better decisions during the planning and execution stages. This capacity not only increases cost accuracy, but it also connects the LCCA with sustainability objectives and performance criteria.

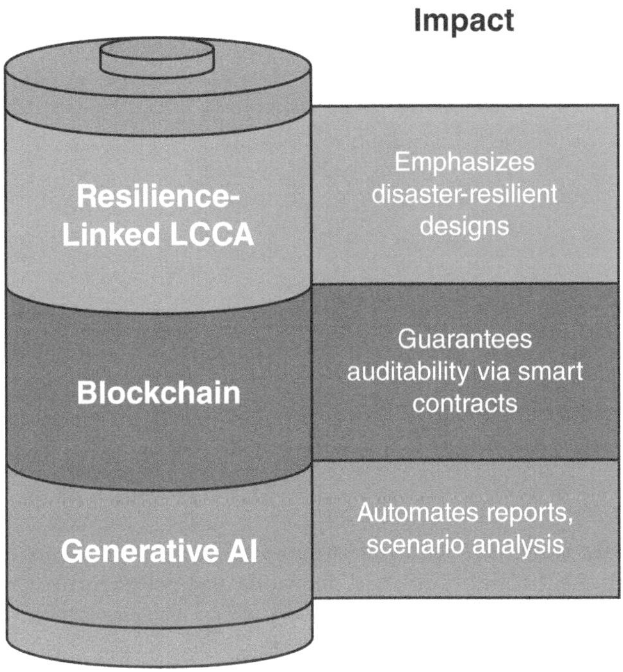

FIGURE 6.5
Innovation in LCCA.

6.14.2 Blockchain for Transparency

Smart contracts monitor each transaction throughout a project's life cycle (e.g., material acquisitions and maintenance records), guaranteeing auditability. Blockchain technology delivers a decentralized, immutable ledger system capable of tracking and verifying project costs throughout its life cycle. Blockchain improves cost traceability and stakeholder responsibility, ensuring that all financial transactions and alterations are transparent and tamper-proof. This is especially beneficial in large, multi-stakeholder projects where cost overruns, procurement fraud, and change orders can disguise actual LCC. With blockchain, LCCA obtains unprecedented legitimacy and auditability.

6.14.3 LCCA Digitalization

With the integration of advanced technologies with historical methods, digitalization has transformed the construction industry nowadays. Technologies such as cloud-based platforms, augmented reality (AR), and BIM transform the project planning process. As a result, the productivity of a project is improved in every phase of its life. Construction has always gone through several challenges, such as improper resource allocation, schedule delays, cost overruns,

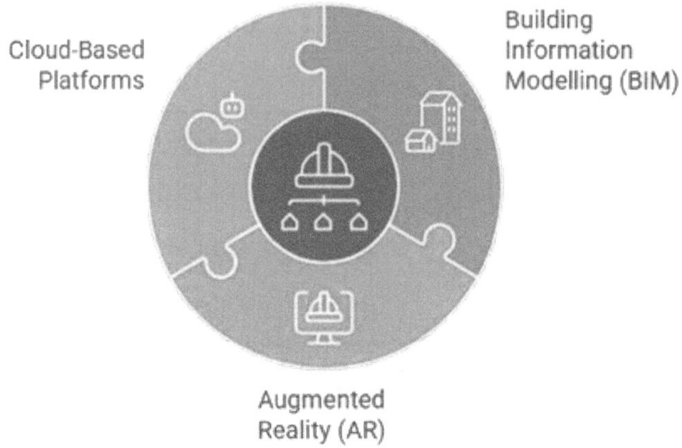

FIGURE 6.6
LCCA digitalization.

and safety. By utilizing digital technologies, stakeholders can overcome all these limitations through informed decisions and better future predictions.

6.14.3.1 Key Tools for Digital Transformation

Figure 6.6 shows three main digital tools in Construction 5.0:

Architects, engineers, and construction professionals can use BIM, a three-dimensional model-based technology, to efficiently plan, design, build, and oversee structures. BIM makes it easier for stakeholders to see architectural details, identify any conflicts, and work together productively.

Construction workers can see projects at different stages, thanks to AR, which superimposes digital information onto the real world. This increases safety procedures, reduces mistakes, and enhances on-site accuracy.

Stakeholders can access project data, monitor progress, and communicate effectively, thanks to cloud technology, which enables real-time collaboration. Decision-making and project transparency are improved by this relationship.

6.15 LCCA in Construction and Renewable Energy Initiatives

LCCA is used to evaluate the total of an entity, including purchasing and operating, maintenance, and disposal costs for its total life. It helps in renewable energy in terms of sustainable and long-term economic viability by validating the energy and maintenance cost of the project. It also supports informed decisions by comparing available options that have the least long-term expenses and at the same time maximum efficiency.

6.15.1 Integration of Digitalization and LCCA

Integration of BIM and a cloud-based platform provides a better methodology in LCCA for planning and executing the project. The life cycle performance of a building and its energy systems can be simulated in BIM. Cloud-based technologies give real-time data so that a manager can make well-informed and updated decisions. This synergy improves cost estimation, better resource utilization, and provides a platform for better decision-making. This process can also be seen in Figure 6.7.

6.15.2 Future of Digitalized LCCA in Smart Construction

The future of construction is indisputably digital. By integrating sophisticated technology with cost analysis approaches such as LCCA, the industry can attain enhanced efficiency, sustainability, and economic resilience. Highlighting long-term planning and real-time data integration, digitalized LCCA is set to be a fundamental element in the advancement of smart construction.

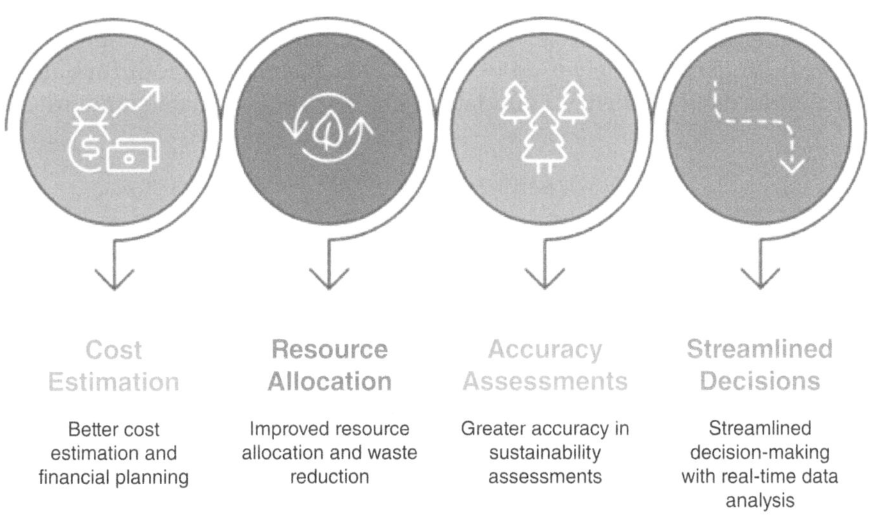

FIGURE 6.7
Benefits of digitalization and LCCA synergy.

6.16 LCCA and BIM

The process for the scientific economic evaluation of LCC is known as LCCA, also known as life cycle costing (International Organization for Standardization [ISO], 2017). The total expenses incurred during each stage of a building's life cycle is known as the LCC. LCC was separated into four categories by ISO 15686-5 (ISO, 2017): construction, operation, maintenance, and end-of-life costs. Whole life cost (WLC), derived from LCC, encompasses all significant initial and recurring expenses and advantages associated with an asset, including LCC, non-construction costs, revenue, and externalities (ISO, 2017). The LCC border is adaptable and may be adjusted according to the prevailing circumstances. BIM holds significant importance in the context of sustainability application. Figure 6.8 illustrates the progressive advancement of BIM in sustainability.

Numerous real-world challenges have prevented LCCA from being widely used from the start. The theory and practice of LCCA have diverged significantly. To put it another way, although being widely accepted, the LCCA concept has not been applied extensively in engineering practice. The primary cause is that the LCCA procedure takes a great deal of time and work (Jansen et al., 2020). The primary advantage of adopting BIM is frequently seen to be the reduction of time and human error, which makes LCCA stable and comfortable for use. Additionally, because BIM can automatically generate the bill of quantities (quantity take-off) used in structures, it can significantly aid in the automatic quantification of required inputs for LCCA. Additionally, BIM-based energy consumption simulation aids in operational cost estimation. These results demonstrate how beneficial BIM is to LCCA (Lu et al., 2023).

BuildingSMART (2021), an international OpenBIM organization, is the source of the most widely used BIM standards. It consists of four fundamental standards: (1) Industry Foundation Classes (IFC) for data standards; (2)

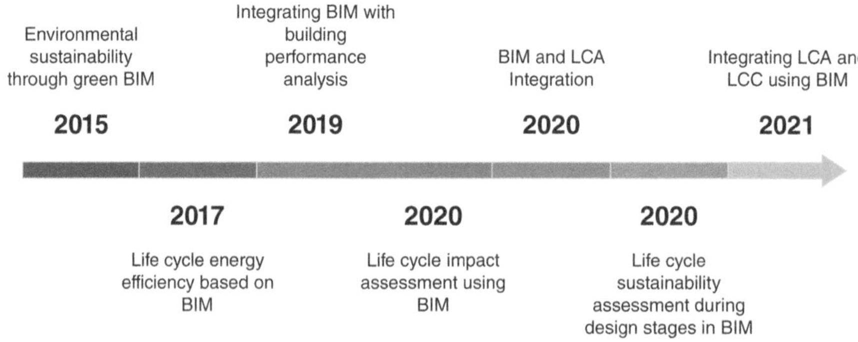

FIGURE 6.8
Evolution of BIM and sustainability.

TABLE 6.4

BIM Uses in Building Life Cycle

Year	Specialization	Finance	Environmental	Public	Author
2015	Environmental sustainability through green BIM		✓		Wong & Zhou, 2015
2017	Life cycle energy efficiency based on BIM		✓		Eleftheriadis et al., 2017
2019	Integrating BIM with building performance analysis	✓	✓	✓	Jin et al., 2019
2020	Life cycle impact assessment using BIM		✓		Crippa et al., 2020
2020	BIM and LCA integration		✓		Potrč Obrecht et al., 2020
2020	Life cycle sustainability assessment during design stages in BIM	✓	✓	✓	LLatas et al., 2022
2021	Integrating LCA and LCC using BIM		✓		Lu et al., 2021

Information Delivery Manual (IDM) for process standards; (3) International Framework for Dictionaries (IFD) for standard lexicons; and (4) Model View Definition (MVD) for process translation. Furthermore, a well-liked standard for life cycle coding is Omniclass (Construction Specifications Institute [CSI], 2021), which combines face and line categorization. BIM has been widely used in the life cycle management of buildings, as shown in Table 6.4.

Benefits of BIM and LCCA Integration

- Providing up-to-date assistance for rapidly forecasting LCC results
- Cutting down on needless building expenses during the design stage
- Promptly upgrading building data to acquire fresh LCC results
- Offering visual representations for engineers to achieve more intuitive outcomes
- Implementing a clear and systematic valuation procedure for cost audits
- Facilitating collaborative efforts within a diverse team

The outcome of the LCC calculation should be presented in an understandable manner. The results of LCC were described in most studies using tables, and many researchers employed pie charts, bar charts, and other charts to vividly depict the results. Furthermore, research has created visually represented BIM-integrated modules to communicate LCC results back to the BIM model, making the results understandable. Numerous prior attempts have been documented. Table 6.5 presents earlier studies on LCCA, detailing

TABLE 6.5

Previous Studies on LCCA

Citation	NPV	BIM Integration	Procedure for Data Exchange	Focus	Further Analysis	Examples
Shin & Cho, 2015	NPV: discount rate	BIM: Graphisoft ArchiCAD BEM: EcoDesigner	Graphisoft ArchiCAD to Microsoft Excel	Choose a suitable design option while taking the LCA and LCCA into account	Sensitivity analysis and environmental assessment (carbon emissions)	A 40-year-old, 11-storey office skyscraper in the Republic of Korea
Ahmad & Thaheem, 2018	NPV: interest rate (0.0418, 0.0746)	BIM: Autodesk Revit	Autodesk Revit to Microsoft Excel	Evaluation of economic sustainability	Economic evaluation: cost-effectiveness, controllability, and flexibility	Standard, ideal and subject, three buildings with 40 years of life
Sandberg et al., 2019	NPV: interest rate BIM	Autodesk Revit, Graphisoft ArchiCAD	Autodesk Revit to Rhino using Grasshopper	Life cycle energy and cost optimization	Environmental evaluation: life cycle energy	A 50-year-old multifamily residential building in Sweden with five and six levels
Ansah et al., 2020	SPW BIM	Autodesk Revit BEM: IES VE	Autodesk Revit to Microsoft Excel	Various facade systems that take LCC and LCA into account	Assessment of environment	A 50-year-old, one-storey structure in Ghana
Tushar et al., 2022	NPV: discount rate (7%)	BIM: Autodesk Revit BEM: FirstRate5	Autodesk Revit to Microsoft Excel	Energy modelling for window systems that combine LCC and LCA	Evaluation of the environment: energy usage	60-year-old window system in Australia

the strategies employed, the methodologies utilized, and the case studies to which they were applied.

6.16.1 Value Engineering

Cost management does not imply that reduced costs are inherently superior, but rather acknowledges the intrinsic worth of expenditures. The implementation of value engineering can guarantee the attainment of LCCA-optimized projects. The investment in home construction or renovation will be offset by decreased energy expenses throughout the life cycle, making value for money a crucial consideration from the inception of a building project in LCCA. The Society of American Value Engineers (SAVE) describes value engineering as the methodical application of established procedures, the determination of a function's worth, and the dependable provision of the requisite function at the minimal total cost. Equation (6.2) denotes the computation of value. The value engineering job plan consists of pre-study, information, creative, assessment, development, presentation, and post-study phases.

$$\text{Value} = \text{Function}/\text{Cost} \tag{6.2}$$

6.16.2 Optimization Measure

Implementing optimization strategies is essential to attain cost-effectiveness by minimizing the LCC and enhancing building functionality. Several variable options influence the structures' LCC outcomes, including facade types, window-to-wall ratios, structural systems, glazing kinds, and building orientation. Utilizing BIM, designers may swiftly obtain LCC outcomes across many schemes and subsequently select the ideal option. The selection of a scheme is particularly crucial for green buildings, low-income structures, and open buildings within the context of value engineering.

6.16.3 Parameter Analysis

The parameters of LCCA may be predicated on erroneous assumptions by owners or technicians, significantly impacting the reliability of value engineering. Consequently, LCC results require uncertainty and sensitivity analysis that accounts for uncertainties and sensitivities arising from the database, risk factors, and parameter configurations. In the uncertainty analysis of LCC, Monte Carlo simulation is the predominant method, although fuzzy sets, Bayesian approaches, and other alternatives may also be viable options. Furthermore, the sensitivity analysis of lifespan or discount rate aids decision-makers in comprehending the economic significance of time. BIM can rapidly update LCC outcomes for value engineering, serving as an efficient tool for parameter analysis.

THOUGHT BOX

Do you consider cost when making design decisions? BIM predicts a structure's financial future in addition to visualizing it. It becomes a powerful tool for decision-making when combined with LCCA.

Is change recognized by your spreadsheet? Spreadsheets are not flexible. Digital platforms are able to. LCCA needs to embrace shared data, interactive models, and ongoing upgrades instead of relying just on static tools.

6.17 Integration of LCCA with Other Technologies

There are some limitations with integration of LCCA with BIM; to overcome these limitations, LCCA can be used with integrating some other technologies, which are shown in Figure 6.9.

6.17.1 OpenBIM

There are problems with compatibility of different BIM tools. Various models are separated because different BIM software providers utilize different storage formats. Furthermore, because the model information from BIM cannot always be transferred automatically, the LCCA programme usually

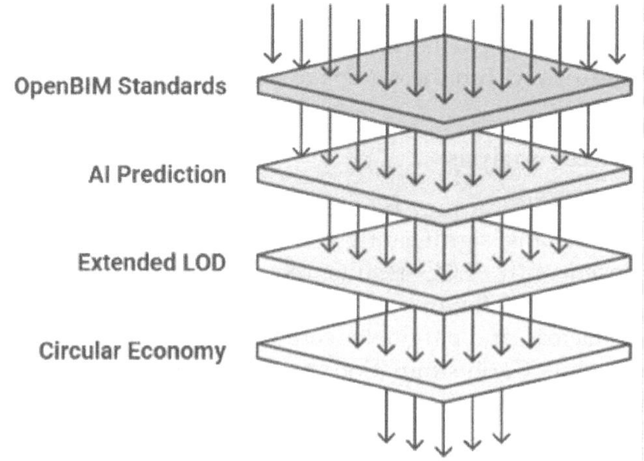

FIGURE 6.9
Other techniques in LCCA.

needs human input. In other words, existing BIM-based LCCA methods for buildings rely on software, and information cannot be transferred or shared between these programmes. This problem might be resolved by the OpenBIM idea, which is defined by consistent data standards. The LCCA can distance itself from software providers, thanks to it. The public BIM data storage standard, IFC, can be used to automatically generate the embedded cost bill of goods. Moreover, IFC can be used to estimate the operational energy usage for cost analysis.

6.17.2 Artificial Intelligence

Existing BIM-based LCCA approaches cannot be used without accurate and comprehensive models. Existing techniques are unable to execute LCC, particularly when it comes to old buildings that were created as built. 3D point cloud scanning can be used to obtain BIM models of older structures, and current BIM-based techniques can be used to solve LCCA. However, this technology cannot be generalized due to the comparatively high expenses of 3D point cloud scanning. Furthermore, issues lacking accurate drawings during the conceptual design stage still cannot be solved with this technique. Machine learning, convolutional neural networks, and artificial neural networks are examples of AI models that may offer a means of predicting and assessing LCC. Although the absence of actual historical data on LCC gathered from prior projects remains a significant practical obstacle, BIM can provide data for LCC estimate using AI. All things considered, by compiling historical LCC data utilizing big data by BIM and a simulation model by AI, owners or consultants can rapidly forecast the LCC performance of new or existing structures.

6.17.3 IoT and Extended LOD

LCC simulation in the design phase, that is, the comparatively early stage of value engineering, is the primary focus of the current study. Value engineering, on the other hand, includes the post-study phase, which is not investigated at this time but may be in the future. Only up until the buildings' as-built phase can LOD500 be declared for BIM accuracy management; after that, it is not expressed. Future research must continue to perform LCCA under enlarged LOD to satisfy post-evaluation requirements for value engineering, even if several studies have tried to extend the LOD to the operation, maintenance, and end-of-life stages.

6.17.4 Circular Economy

The majority of the BIM-based LCCA patterns found in previous research are linear economies. The LCC results in these studies are obtained by linearly aggregating the resources from construction, operation, and destruction.

Many building resources and components retain their worth long after a building's service life is over. Economic sustainability can be attained if the financial impact of component recycling is considered. According to the circular economy concept, building components and resources must be used as efficiently as possible in order to maintain a continuous cycle of use, reuse, repair, and recycling. Additionally, BIM can document component and resource information, which will be useful for LCCA in future studies under the circular economy.

THOUGHT BOX

Will the LCCA of the future be climate-proofed, chain-stored, and constructed by machines? Resilience metrics, generative AI, and blockchain are the future of life cycle intelligence, not fads.

6.18 Intelligent Construction

Intelligent construction denotes a revolutionary shift in the design, implementation, and administration of construction projects. It integrates advanced technologies like AI, BIM, the IoT, robotics, and data analytics to enhance the overall life cycle of a project. The principal objective is to improve efficiency, reduce costs, minimize human errors, and strengthen sustainability. The McKinsey Global Institute's 2017 research 'Reinventing Construction: A Route to Higher Productivity' reveals that the construction industry has experienced productivity stagnation for decades, unlike significant progress in other sectors. This shift towards advanced construction methodology aims to address these inefficiencies through the adoption of sophisticated digital tools and technologies that transform traditional construction processes. The construction sector is undergoing a substantial transformation propelled by the rapid evolution of technology. The integration of AI into building methods is a revolutionary change that is altering the design, execution, and management of projects. At PMCBS, are spearheading this change by leveraging AI to improve efficiency, precision, and creativity in construction management. This newsletter explores the revolutionary potential of AI in construction, its essential role, the advantages it offers, and the promising future trends in AI integration within the industry.

6.18.1 What Is the Rationale behind Intelligent Construction?

About 90% of megaprojects surpass their budgetary allocations (Oxford University, 2021). Buildings account for 39% of worldwide CO_2 emissions

TABLE 6.6
Global Adoption of Intelligent Construction

Country	Key Initiative	Impact
China	'Intelligent Construction 2035' plan	$500 billion investment in robotic construction
USA	DARPA's 'Automated Construction' program	AI-driven military bases in <30 days
UAE	Dubai's blockchain construction permits	Reduced approval times from 30 days to 1 hour
Germany	Autobahn digital twin	Predictive maintenance saves €1 billion/year

(UNEP, 2022). Global Reports Supporting IC Adoption McKinsey (2023): 'Reinventing Construction through a Productivity Revolution' anticipates $1.6 trillion in yearly savings from digital transformation. World Economic Forum (2022): 'The Future of Construction' emphasizes modular construction and AI as primary disruptors. Table 6.6 shows the global adoption of intelligent construction with the key initiatives taken and their impact.

6.19 Advancements in Smart Construction Technologies

6.19.1 Building Information Modelling

BIM has transformed the design and construction processes by facilitating the digital depiction of the physical and functional attributes of facilities. This 3D model facilitates successful collaboration among all stakeholders, including architects, engineers, contractors, and owners. BIM facilitates simulations and real-time project oversight, reducing errors and enhancing resource optimization. The Skanska 2018 report on BIM indicated that the adoption of BIM resulted in a 40% reduction in errors and a 20% decrease in project expenses. As BIM advances, 5D BIM, which incorporates cost data, and 6D BIM, which includes sustainability elements and lifetime data, are set to become industry benchmarks. The future of BIM will likely integrate more immersive technologies, such as AR and virtual reality (VR), to enhance representations.

6.19.2 Internet of Things

To provide real-time data of construction sites, the IoT is considered a crucial technology. Integrating sensors into building materials, machinery, and equipment allows construction managers to oversee site conditions, monitor equipment usage, and ensure safety. This technology enables predictive maintenance, reducing unexpected downtimes and extending asset lifespan. The worldwide Construction 2030 Report anticipates substantial expansion in the worldwide construction sector, propelled by advancements in IoT and other digital technologies. It forecasts that IoT will enable the development of progressively 'connected' construction sites, where equipment, humans, and materials interact seamlessly.

In Singapore, the BCA Smart Urban Rail initiative employed IoT to monitor real-time data for the development of its metro line. Sensors installed on construction machinery and within building materials facilitated the monitoring of structural integrity and environmental conditions, which were essential for the project's safety and efficiency.

THOUGHT BOX

Is your construction methodology aligned with your technological advancements? Intelligent construction is not merely a trend; it is an imperative. Given the industry's escalating prices, stringent deadlines, and increased sustainability requirements, the utilization of AI, BIM, IoT, and robotics is becoming imperative rather than discretionary.

6.20 The Function of Artificial Intelligence in Construction

AI and machine learning are set to transform decision-making in the construction industry. AI algorithms can evaluate extensive historical and real-time data to forecast project outcomes, including scheduling delays, cost overruns, and safety risks. AI is utilized in automated design processes, enhancing the efficiency and accuracy of architectural designs. AI assumes a diverse role in the building sector. Key domains in which AI exerts a substantial influence are shown in Figure 6.10.

6.20.1 Project Planning and Design

AI algorithms can analyse historical data and current trends to improve project planning and design. It encompasses the creation of complex 3D models, anticipating future issues and proposing solutions before the initiation of construction.

6.20.2 Site Surveillance and Safety

AI-powered drones and cameras deliver real-time site monitoring, ensuring compliance with safety requirements and identifying potential risks.

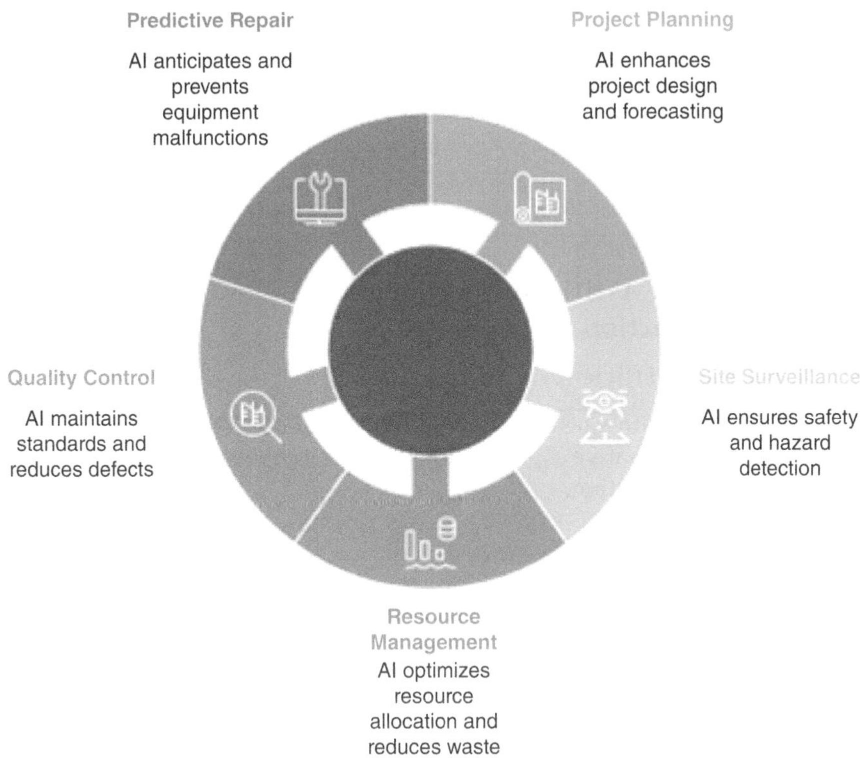

FIGURE 6.10
AI domains in construction.

Machine learning models can predict and prevent accidents by analysing patterns and providing safety recommendations.

6.20.3 Resource Management

AI enhances resource allocation efficiency by predicting material requirements, labour needs, and equipment usage. It minimizes waste and ensures the availability of resources as needed.

6.20.4 Quality Control

AI systems can assess and monitor construction quality through image recognition and data analysis. This ensures compliance with the necessary standards and reduces the likelihood of faults.

6.20.5 Predictive Repair

AI can foresee equipment failures and coordinate repairs, thus preventing costly disruptions and extending the operating lifespan of machines.

6.21 Benefits of Artificial Intelligence in Construction Management

6.21.1 Increased Efficiency

AI automates repetitive tasks, allowing human workers to focus on the project's complex and creative aspects. This leads to accelerated completion times and improved production.

6.21.2 Enhanced Accuracy

AI reduces the probability of human error by providing precise data analysis and predictions. It ensures that projects are conducted to the highest standards.

6.21.3 Cost Reduction

By optimizing resources and predicting issues, AI aids in decreasing overall project costs. Efficient project management and proactive maintenance further improve cost efficiency.

6.21.4 Improved Safety

AI enhances site safety through real-time monitoring and predictive analytics. This reduces the likelihood of accidents and ensures a safer working environment.

6.21.5 Data-Driven Decisions

AI offers critical insights via data analysis, facilitating informed decision-making. It results in improved project outcomes and enhances strategic planning. The Deloitte 2020 Construction Industry Outlook indicates that AI can enhance construction schedules, resource distribution, and budget predictions. It can automate processes like material procurement, minimizing human involvement and decreasing the likelihood of errors. Instruments such as ALICE Technologies and nPlan forecast delays and enhance schedules with 90% precision (Forbes, 2023).

Case Study: The LA Metro expansion utilized AI to decrease scheduling conflicts by 75%, resulting in savings of $200 million.

6.22 Drones and Unmanned Aerial Vehicles

Drones have become essential instruments on building sites. Although this is not groundbreaking news, their established roles in military, economic, and recreational contexts render their impact on building considerable. These unmanned aerial vehicles (UAVs), especially the autonomous variants, function without direct human oversight, thereby integrating them into the robotic workforce. Drones in construction are often classified into five primary categories: site study, planning and design, asset inventory, project reporting and cooperation, and dispute settlement. These UAVs deliver real-time data and optimize multiple facets of construction, facilitating project management and ensuring adherence to timelines. For the improvement in efficiency and safety, drones have come up as a vital tool. Mapping and land surveys are one of the most common examples of its usage. With the help of high-resolution cameras and LiDAR sensors, drones can generate detailed 2D as well as 3D maps of an area. Through this efficient work, surpassing traditional survey methods saves time and cost, and provides real-time information to design and construction teams. This data can be integrated with BIM, which further facilitates real-time updates of the planned model, hence improving the overall efficiency of a project. Drones can also be used for inventory management by scanning QR codes to recognize specific material and ensuring accurate inventory, ultimately lowering the chances of loss and theft.

For overseeing the progress of projects, drones are considered crucial. They provide a detailed visual of construction sites by conducting regular flights, so that managers can take swift actions and monitor the progress effectively. When drones are integrated with AI, they can be used to predict potential issues and reduce project delays. They also improve the safety of various structures like skyscrapers, bridges, or any confined place by surveying unreachable areas, which also reduces the effort of manual inspection. Drones further enhance occupational safety. They can be employed to monitor compliance with safety requirements, including the proper use of personal protective equipment (PPE), and to detect dangerous behaviours or conditions in real time. Thermal drones can identify heat stress and many health risks. Drones enhance the creation and maintenance of digital twins' virtual representations of real assets by collecting aerial and structural data for simulation, remote inspections, and predictive maintenance. Drones are being employed for site security. They can autonomously oversee construction sites, particularly during non-operational hours, to prevent unauthorized access and theft. When integrated with AI-driven analytics, drones can detect motion, recognize faces, and deliver real-time alerts to security personnel. Drones are essential to smart construction, offering a powerful combination of real-time data collection, automation, and integration with digital technologies like BIM, IoT, and AI. Their application enhances project efficiency, safety, and decision-making across the whole construction life cycle.

6.23 Generative AI for Immediate Blueprints

Platforms such as OpenAI's Codex and Autodesk will enable engineers to articulate a building in natural language, prompting AI to provide numerous optimum designs within minutes. The UK Crossrail initiative employed AI to enhance construction timelines and resource distribution. AI was employed to analyse extensive datasets from building activities, facilitating improved management of project timetables and expenditures.

6.24 Robotics and Automation

Robotics is revolutionizing labour-intensive activities, like bricklaying, welding, and building inspection. Drones and robotic arms are progressively employed for activities like site surveys and the assembly of prefabricated components, markedly accelerating building operations.

- **Hyper-Automation and Robotics**
  By 2030, 30–50% of building activities will be automated (McKinsey, 2023). Significant progress encompasses:
- **Self-Operating Heavy Equipment**
  Firms such as Built Robotics and Komatsu are implementing autonomous excavators and bulldozers that function continuously, achieving 10–20% more efficiency.
  Case Study: A Swedish highway project employed autonomous Volvo dump trucks, resulting in a 15% reduction in fuel expenses and a 90% decrease in accidents.
- **Swarm Robotics for Extensive Initiatives**

Collaborative robot teams will operate in unison some constructing bricks, others welding steel, all coordinated through 5G and AI. ETH Zurich's DFAB House was constructed with many robotic technologies in coordination, reducing building time by 40%.

Case Study: A 2021 research by the World Economic Forum emphasized that robotics in construction is transformative, forecasting that automation will comprise 50% of construction jobs by 2040. Robotics not only augments productivity, but also bolsters safety by executing hazardous activities often managed by human labourers.

The Shenzhen International Airport in China utilized robotic technology for construction, featuring a bricklaying robot capable of laying up to 1,000 bricks per hour, significantly enhancing building speed while maintaining accuracy.

6.25 Augmented Reality and Virtual Reality

AR and VR technologies will experience extensive utilization in construction for design visualization, virtual walkthroughs, and improved project planning. These technologies enable clients, architects, and contractors to envision the project in three dimensions or engage with it in real time. The 2020 VR and AR Construction Market Report indicated that the valuation of VR and AR technologies in construction reached $1.1 billion in 2020, with an anticipated compound annual growth rate (CAGR) of 34.6%. These tools facilitate enhanced communication and decision-making inside teams by offering immersive, data-driven project visualizations.

In Finland, the VTT Technical Research Centre employed AR to facilitate the development of a smart building. Employees donned AR glasses, which supplied them with instantaneous data and visual enhancements of the building schematics. The incorporation of AR resulted in a 15% decrease in errors and a 10% enhancement in project efficiency.

6.26 Data-Driven Decision-Making and Predictive Analytics

The future of intelligent construction relies on data-informed decision-making. As construction projects produce an increasing volume of data, predictive analytics will enable stakeholders to make more informed decisions. Through the integration of AI and big data, construction managers may proactively identify issues such as delays, cost overruns, and safety dangers prior to their manifestation.

The 2019 PwC research titled *The Future of Construction* indicated that construction firms employing data-driven methodologies had a 30% enhancement in decision-making velocity and a 20% decrease in expenses contributed to improved resource management.

The Sydney Opera House employed predictive analytics to evaluate data from prior projects, resulting in a 25% reduction in material waste.

The project team utilized predictive data to enhance resource utilization and minimize delays, illustrating the efficacy of data-driven construction management.

6.27 Sustainability and Eco-Friendly Construction Technologies

Sustainability is a fundamental priority for the future of construction. The construction sector is a significant contributor to global carbon emissions; nevertheless, the emergence of smart construction technology enables the industry to reduce its environmental impact. Green building technologies, encompassing energy-efficient systems, renewable energy integration, and sustainable materials, are pivotal to this shift.

The Green Building Council's 2020 study on sustainable construction forecasts that green building technologies would attain a global market value of $574 billion by 2022, facilitated by smart technologies that enhance energy efficiency and optimize materials management.

The Edge in Amsterdam, frequently referred to as the 'most intelligent office building globally', employs renewable energy, advanced lighting systems, and IoT sensors to track and minimize energy consumption. These advances have resulted in a 70% decrease in energy consumption compared to traditional office buildings.

6.28 Enhanced Project Management Using AI and Cloud-Based Collaboration

Intelligent construction will fundamentally transform project management. AI-driven project management solutions will automate numerous facets of construction administration, encompassing scheduling, procurement, and budgeting. Cloud-based solutions will improve cooperation, enabling teams to access current information and make real-time choices. The 2019 Smart Construction Market research indicates that cloud-based solutions are projected to expand by 40% over the next five years, facilitating enhanced collaboration and decision-making. The Burj Khalifa, the tallest edifice globally located in Dubai, employed a cloud-based project management system to enhance communication among its teams and monitor materials, timelines, and financial allocations. This real-time collaboration was crucial for the timely completion of the project.

THOUGHT BOX

Why compromise human safety when machines are capable of flying, scanning, and constructing? Drones and autonomous robots mitigate risks, enhance accuracy, and optimize processes. They not only ascend but also assume accountability for safer, more intelligent locations.

6.29 Anticipating Future Outcomes with Artificial Intelligence

6.29.1 Predictive Analytics

Construction managers utilize historical data like previous projects' cost, schedules, and safety incidents, and utilize AI to predict future trends. Construction professionals may reduce the risk associated with every situation they face by utilizing predictive models that incorporate statistical modelling and machine learning. These technologies improve budgeting precision and realistic project timelines, hence increasing overall project efficiency and reducing potential cost overruns.

6.29.2 Anticipated AI-Driven Technologies

Across the construction industry, AI-driven technologies are being applied to different applications, for example utilizing Artificial Neural Network (ANN) for cost overrun predictions and other learning techniques to make better planning and decisions. Also, these AI solutions improved the performance of buildings in terms of lower energy usage and better indoor air quality. Implementation of these technologies, especially at the initial stages for feasible studies and assessment, is further increasing their importance and planning a project. These milestone achievements are useful for better performance and efficiency. Also, it suggests better future performances. As we go into a future where construction approaches incorporate advanced AI technologies, we foresee reduced costs, expedited project timelines, and substantial improvements in construction quality. The consequences of these advancements extend beyond immediate gains in efficiency and profitability, indicating a more resilient, adaptable, and innovation-driven industry. The predictive powers of AI and its impact on improving labour skills are revolutionizing the construction industry, making it more efficient and increasingly compatible with sustainability and advanced project management demands. This expedition, directed by AI, guarantees a physical environment crafted not just for the now but also with a distinct emphasis on the future.

A report Procore 2024 was published by PROCORE United States by conducting 1,200 international experts' surveys. This report suggests the following findings for the future of smart construction:

- AI and robotics are set to substantially improve productivity in the construction sector. Currently, 18% of project time is expended on data retrieval, while 28% is squandered on rework.
- About 55% of construction executives anticipate that automation will transform the sector within the next five years. Integrating AI-driven preconstruction technologies and automation enables firms to diminish inefficiencies, enhance project planning, and increase ROI.
- The construction workforce is experiencing a generational shift. As 53% of employees are projected to retire by 2036, organizations are implementing measures to respond to this transition.
- About 47% of organizations now have upskilling programmes, while 41% intend to establish one during the next year. These initiatives are essential for attracting and maintaining personnel, particularly as 60% of workers remain in construction for professional advancement and financial prospects.
- Furthermore, 45% of employees identify burnout as the primary cause for their departure from the sector, underscoring the significance of mental health assistance and well-being programmes.
- More than 80% of construction executives concur that integrated historical data is essential for project success, whereas 76% of civil and infrastructure builders admit they have not yet fully leveraged their data's potential.
- AI-generated insights will enhance tasks such as scheduling and labour distribution, while human experience will continue to be crucial in the efficient application of these insights.
- Digital tools enabling builders to spearhead the design process. About 49% of surveyed construction professionals expect heightened utilization of BIM for design collaboration and clash identification.
- The design process is transforming as builders assume greater control. Digital twins and AI-driven simulations will enable teams to refine designs prior to the commencement of construction.

According to the Digitalisation in Construction Report 2024 (RICS, November 2024), survey responses from 2023 indicate that digital technology is mainly used to improve existing processes such as cost estimation, forecasting, planning, project management, and progress tracking, including health, safety, and well-being. In contrast, areas like environmental, social, and governance (ESG) standards, full life-cycle and asset management, social value, and carbon tracking see less digital adoption. Overall, digital technologies and data exchange play a crucial role in the construction and retrofitting of buildings and infrastructure that deliver social benefits. The 2023 survey responses reveal that digital technology is primarily utilized to enhance existing procedures, including cost estimation, forecasting, planning, management, and enhancement of progress tracking, health, safety, and well-being. In contrast, areas such as environmental, social, and governance (ESG) standards, full life cycle and asset management, social value, and carbon tracking are experiencing less involvement. Digital technologies and data exchange are essential in the construction and retrofitting of buildings and infrastructure that provide social benefit.

6.30 Conclusion

The integration of LCCA with modern digital technology represents a significant step in the construction industry's quest for sustainability, efficiency, and long-term economic viability. The sector, previously limited by short-term financial criteria, is now shifting towards a comprehensive perspective on project value that encompasses the total cost of ownership, environmental impact, and future adaptability. Digitalized construction facilitated by AI, BIM, IoT, drones, and robotics has made major improvements in the design as well as in the execution phase of the construction project. All these technologies have been merged to make better decisions and to improve the overall life cycle of the project. The integration of advanced construction techniques and LCCA could help to make real-time adjustments, predict the future, and resilient design, considering economic, environmental, social, and laws and regulatory factors. This will improve overall efficiency, productivity, reduce carbon footprints, and improve safety standards

The shift from static spreadsheets to dynamic, intelligent models signifies a future where buildings are continuously optimized rather than simply constructed. In the post-Industry 5.0 era, the construction industry must embrace digitization, circular economy principles, and creative cost strategies. LCCA, enhanced by new technology, functions not just as a tool, but also as the foundation for creating more intelligent, sustainable, and economically resilient environment for future generations.

Chapter Review: Chapter 6 – Life Cycle Cost Analysis in Construction Projects and the Future of Intelligent Construction

Multiple Choice Questions

1. What is the primary purpose of Life Cycle Cost Analysis (LCCA)?
   A. Estimating project duration
   B. Minimizing initial construction cost
   C. Evaluating total cost of ownership over time
   D. Assessing construction safety
2. Which of the following cost components is unique to LCCA compared to traditional cost analysis?
   A. Capital cost
   B. Operational cost
   C. Maintenance cost
   D. Residual value
3. What is typically excluded in Net Present Value (NPV) calculations for LCCA?
   A. Future costs
   B. Discount rate
   C. Inflation rate
   D. Emotional impact of investment
4. Which technology allows stakeholders to track costs transparently throughout a project's life cycle?
   A. BIM
   B. Blockchain
   C. Generative AI
   D. Cloud storage
5. How does resilience-linked LCCA differ from conventional models?
   A. It ignores future costs.
   B. It only considers energy savings.
   C. It incorporates climate risk and adaptability.
   D. It replaces cost calculations with AI predictions.
6. Which of the following is a benefit of integrating BIM with LCCA?
   A. Increased physical labour
   B. Higher insurance premiums

   C. Automated cost forecasting and visualizations
   D. Reduction in building lifespan
7. What role does artificial intelligence play in LCCA?
   A. Removing uncertainty entirely
   B. Predicting maintenance and cost trends
   C. Designing HVAC systems
   D. Blocking user access to financial models
8. Which of the following is an example of a recurring cost in LCCA?
   A. Land acquisition
   B. HVAC servicing
   C. Structural demolition
   D. Design consultant fees
9. Which of the following most accurately defines the primary aim of Intelligent Construction?
   A. To expand the workforce of construction personnel on site
   B. To shorten project schedules via manual labour
   C. To incorporate advanced technology to improve efficiency and sustainability
   D. To depend exclusively on conventional construction materials and methods
10. Which innovation is employed in smart construction to dynamically modify cost models utilizing real-time data?
   A. Accelerators for concrete curing
   B. Dynamic LCCA models associated with digital twins
   C. Conventional spreadsheets and manual records
   D. Manual cost estimation charts

Answer Key

Question No.	Correct Option
1	C
2	D
3	D
4	B
5	C

Question No.	Correct Option
6	C
7	B
8	B
9	C
10	B

References

Ahmad, T., & Thaheem, M. J. (2018). Economic sustainability assessment of residential buildings: A dedicated assessment framework and implications for BIM. *Sustainable Cities and Society, 38*, 476–491.

Ahmed, I. M., & Tsavdaridis, K. D. (2018). Life cycle assessment (LCA) and cost (LCC) studies of lightweight composite flooring systems. *Journal of Building Engineering, 20*, 624–633.

Allison, J. A. (2012). The financial crisis and the free market cure. *Reason Papers, 34*(2), 229–233.

Ansah, M. K., Chen, X., Yang, H., Lu, L., & Lam, P. T. (2020). An integrated life cycle assessment of different façade systems for a typical residential building in Ghana. *Sustainable Cities and Society, 53*, 101974.

Cabeza, L. F., Rincón, L., Vilariño, V., Pérez, G., & Castell, A. (2014). Life cycle assessment (LCA) and life cycle energy analysis (LCEA) of buildings and the building sector: A review. *Renewable and Sustainable Energy Reviews, 29*, 394–416.

Crippa, J., Araujo, A. M., Bem, D., Ugaya, C. M., & Scheer, S. (2020). A systematic review of BIM usage for life cycle impact assessment. *Built Environment Project and Asset Management, 10*(4), 603–618.

Dell'Isola, A. (1997). Value engineering. *Practical applications… for design, construction, maintenance & operations*. RS Means Company Inc.: Kingston MA.

Dwaikat, L. N., & Ali, K. N. (2018). Green buildings life cycle cost analysis and life cycle budget development: Practical applications. *Journal of Building Engineering, 18*, 303–311.

Eleftheriadis, S., Mumovic, D., & Greening, P. (2017). Life cycle energy efficiency in building structures: A review of current developments and future outlooks based on BIM capabilities. *Renewable and Sustainable Energy Reviews, 67*, 811–825.

Jansen, B. W., van Stijn, A., Gruis, V., & van Bortel, G. (2020). A circular economy life cycle costing model (CE-LCC) for building components. *Resources, Conservation and Recycling, 161*, 104857.

Jin, R., Zhong, B., Ma, L., Hashemi, A., & Ding, L. (2019). Integrating BIM with building performance analysis in project life-cycle. *Automation in Construction, 106*, 102861.

Lakousha, M. M. (2023). Con-LCCA V1. 0: A computerized tool for analyzing the life cycle cost of construction projects. *SVU-International Journal of Engineering Sciences and Applications, 4*(1), 62–78.

Langston, C. (2013). *Life-cost approach to building evaluation*. Routledge.

LLatas, C., Soust-Verdaguer, B., Hollberg, A., Palumbo, E., & Quinones, R. (2022). BIM-based LCSA application in early design stages using IFC. *Automation in Construction, 138*, 104259.

Lu, K., Deng, X., Jiang, X., Cheng, B., & Tam, V. W. (2023). A review on life cycle cost analysis of buildings based on building information modeling. *Journal of Civil Engineering and Management, 29*(3), 268–288.

Lu, K., Jiang, X., Yu, J., Tam, V. W., & Skitmore, M. (2021). Integration of life cycle assessment and life cycle cost using building information modeling: A critical review. *Journal of Cleaner Production, 285*, 125438.

Potrč Obrecht, T., Röck, M., Hoxha, E., & Passer, A. (2020). BIM and LCA integration: A systematic literature review. *Sustainability, 12*(14), 5534.

Sandberg, M., Mukkavaara, J., Shadram, F., & Olofsson, T. (2019). Multidisciplinary optimization of life-cycle energy and cost using a BIM-based master model. *Sustainability, 11*(1), 286.

Shin, Y.-S., & Cho, K. (2015). BIM application to select appropriate design alternative with consideration of LCA and LCCA. *Mathematical Problems in Engineering, 2015*(1), 281640.

Tushar, Q., Bhuiyan, M. A., & Zhang, G. (2022). Energy simulation and modeling for window system: A comparative study of life cycle assessment and life cycle costing. *Journal of Cleaner Production, 330,* 129936.

Wong, J. K. W., & Zhou, J. (2015). Enhancing environmental sustainability over building life cycles through green BIM: A review. *Automation in Construction, 57,* 156–165.

Index

Note: Locators in *italics* represent figures and **bold** indicate tables in the text.